Praise for

BACK TO HUMAN

'Technology may have accelerated the pace of change, but it hasn't erased the need for business basics. In *Back to Human*, Dan Schawbel offers expert advice on overcoming technology's shortcomings and refocusing on the true building blocks for business success: relationships, collaboration, and getting the job done. A must-read for all leaders.'

—Ron Shaich, founder and chairman of Panera Bread

'*Back to Human* shows how modern technologies have made our work lives unfulfilled. In this provocative and insightful book, Dan Schawbel demands our interactions to be more human and less machine, and he provides a practical guide on how to achieve this. This book is a must-read for anyone who wants to become a more effective leader in today's workplace.'

—W. Chan Kim, the BCG professor of strategy at INSEAD and *New York Times* bestselling author of *Blue Ocean Shift*

'Communication is leadership. That's what leadership is: communication. *Back to Human* brilliantly coaches us in successfully making technology and our devices cause us to be better leaders and communicators rather than worse ones.'

—Kip Tindell, cofounder, chairman, and former CEO, The Container Store

'In his brilliantly researched new book, *Back to Human*, Dan Schawbel provides the deepest, most insightful analysis of how we can restore humanity and authentic connections to the workplace in this technology age: with leaders who build highly motivated, collaborative teams that create healthy, productive workplaces. It is a must-read for all who care about making work fulfilling.'

—Bill George, senior fellow, Harvard Business School; former chair and CEO of Medtronic; and author of *Discover Your True North*

'I recommend *Back to Human* to any leader who wants to create a higher quality of life for their team. Schawbel explains how to build the human connections that are critical to personal and organizational success. Regardless of advancements in technology, the human touch will remain, and this book will help you create stronger relationships that lead to higher performance and happiness.'

—Michel Landel, CEO of Sodexo

'Dan has written a meaningful neo classic. He reinforces the human need for fulfillment and shows that most technology limits this important connection. When applied, his ideas will help people find meaningful connections that increase both personal well-being and work productivity.'

—Dave Ulrich, coauthor of the *New York Times* bestseller *The Why of Work*; Rensis Likert professor, Ross School of Business, University of Michigan

'In *Back to Human*, Dan Schawbel reminds us that our humanity must never take a back seat to new technology—driverless or otherwise. And that progress is found in the connection of humans working together creatively in pursuit of better.'

—Beth Comstock, former vice chair, GE

'*Back to Human* is a valuable read for any leader desirous of a more collaborative and productive workforce. By following Dan's advice, we can enjoy stronger team relationships that lead to stronger business results.' —Bert Jacobs, cofounder and chief executive optimist, Life Is Good

'In *Back to Human,* Dan Schawbel challenges us to put down our phones and start investing in deeper relationships. That's a message we all need to hear.'

—Dan Heath, coauthor of the *New York Times* bestsellers *The Power of Moments, Made to Stick, Switch* and *Decisive*

'There are books on productivity that ignore practical lessons, and books with checklists that ignore the 'why.' But what if there was a research-based book chock-full of highly relevant exercises that can help anyone become more effective at work? Look no further: *Back to Human* is loaded with practical insights that won't only help you get better on the job but will also give you pause to think about how to live the life you really want to.'

—Sydney Finkelstein, Dartmouth professor and bestselling author of *Superbosses* and *Why Smart Executives Fail*

'*Back to Human* provides practical back to basics on how to become a better leader. Good insights coupled with good advice. Well done!'

—David Novak, former chairman and CEO of YUM! Brands

'In colonial times, Samuel Adams and his fellow revolutionaries met in taverns and planned the American Revolution over a beer or two. *Back to Human* leads us to recapture those essential human interactions—conversation, communication, collaboration and common passion. If you want to brew up your own revolution, this book by Dan Schawbel is an excellent guide. Cheers!'

—Jim Koch, founder and brewer of Samuel Adams and author of *Quench Your Own Thirst: Business Lessons Learned Over a Beer or Two*

'If leaders want to create stronger connections with their teams, they must read *Back to Human*. Schawbel's message of encouraging more human connection, instead of relying on technology, will become more relevant over time.'

—Howard Behar, former president of Starbucks

'A very timely book on a subject the world needs to pay attention to!'

—Gary Keller, founder of Keller Williams Realty International, *New York Times* bestselling author of *The ONE Thing*

'Technology makes a wonderful servant but a terrible master. Unfortunately, over time it has progressed from the former to the latter and in many ways has created both business and personal havoc. Fortunately, one of the sharpest minds in the business world today, Dan Schawbel, has once again stepped up to the plate and shared his immense wisdom with us! Not only has he identified the problem, he has also systematically provided the solutions to help leaders create a culture and environment in which their team members can enjoy and thrive in their work. This is a book that every leader—that every person—should own, devour and keep by their desk for handy reference.'

—Bob Burg, coauthor of *The Go-Giver* and *The Go-Giver Influencer*

'If you can buy only one book this year to enhance your relationships and build your career or business, *Back to Human* is the one. Grounded in valid research, *Back to Human* is a compendium of strategies, ideas and practical exercises and advice from numerous leaders that work.'

—Susan RoAne, speaker and author of *How to Work a Room* and *Face to Face: How to Reclaim the Personal Touch in a Digital World*

BACK TO HUMAN

Also by Dan Schawbel

Promote Yourself: The New Rules for Career Success

Me 2.0: 4 Steps to Building Your Future

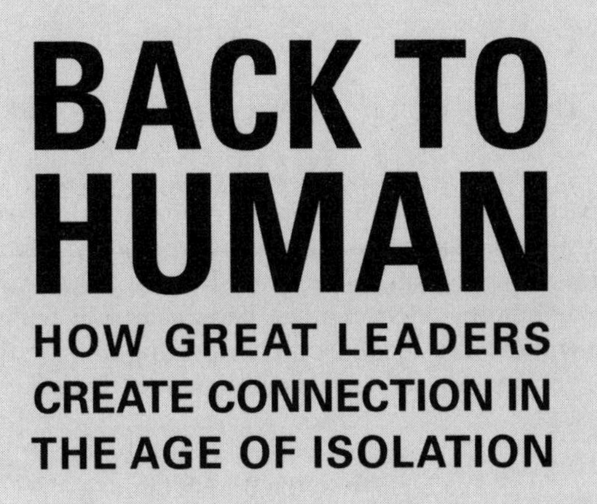

Dan Schawbel

PIATKUS

First published in the US in 2018 by Da Capo Press, an imprint of
Perseus Books, LLC, a subsidiary of Hachette Book Group, Inc.

First published in Great Britain in 2018 by Piatkus

3 5 7 9 10 8 6 4 2

A CIP catalogue record for this book
is available from the British Library.

ISBN 978-0-349-42235-0

Printed and bound in India by Manipal Technologies Limited, Manipal

Papers used by Piatkus are from well-managed forests
and other responsible sources.

Piatkus
An imprint of
Little, Brown Book Group
Carmelite House
50 Victoria Embankment
London EC4Y 0DZ

An Hachette UK Company
www.hachette.co.uk

www.improvementzone.co.uk

To Jim Levine, my literary agent

Contents

Introduction

How Technology Is Isolating Us at Work

Our hyperconnectedness is the snake lurking in our digital Garden of Eden.

—Arianna Huffington[1]

While watching *Black Mirror*, the popular British science fiction anthology, on Netflix, I was amazed at how well the "Nosedive" episode reflects our current society. The story is set in an alternate reality where people can rate each other using their smartphones and those ratings impact their lives. Lacie, the lead character, is obsessed with her rating, much like many of us obsess over the number of likes, comments, and shares our status updates gain. She starts the episode with a rating of 4.2 out of 5 but needs at least a 4.5 to be able to move into the more luxurious neighborhood where her friends live. Her friend Naomi, who has a 4.8 rating, asks Lacie to be a bridesmaid at her wedding. On her way to the wedding Lacie encounters a series of misfortunes that drive her rating down to a mere 2.6. As a result, Naomi asks her not to be in her wedding anymore. Although the show is fictional, this episode perfectly illustrates how technology can divide us as much as bring us together, and it holds up an unforgiving mirror that shows us how guilty we are of making unconscious social comparisons that make us—and everyone else—miserable.

Modern technologies have impacted our workplaces in ways that would have been impossible just a decade ago. Instant messaging, digital platforms, and videoconferencing have completely changed how, when, and where we work. A Gallup survey found that over a third of the entire US workforce has worked remotely, and Freelancers Union reports that freelancers also now make up more than a third.[2] Robotics and artificial intelligence have supercharged our productivity at the cost of replacing tasks and even eliminating full-time jobs from our economy. McKinsey found that half of today's work activities could be automated by 2055, accounting for almost $15 trillion in wages globally.[3]

On the positive side, networks, apps, and smartphones have created a more social, collaborative, and flattering global workplace. According to the *Harvard Business Review*, over the past twenty years collaborative activities have increased by 50 percent and now account for more than 75 percent of an employee's day-to-day work.[4] But more and more of that collaboration is occurring within social networks and mobile apps, with a far smaller percentage happening in person. There's no stopping the evolution of these technologies; they'll continue to transform and reshape our work lives every year.

To give you a sense of just how quickly things are changing, when the telephone was introduced in the latter part of the nineteenth century, it took several decades for that new technology to reach half of all households. A century later, in the 1990s, it took fewer than five years for cell phones to reach the same penetration.[5] Future devices could prove even faster than that.

Our devices offer many incredible benefits, including real-time interactions, efficiencies in workflow, creation of new ideas, and access to resources. At the same time, those devices have disrupted our relationships and made our workplaces more dysfunctional. Instead of strong bonds, we have weak ties. Instead of productive meetings, we have distractions. Technology has created an illusion that today's workers are highly connected to one another, when in reality most feel isolated from their colleagues.

What they crave most—and what research increasingly shows to be the hallmark of the highest performing workplace cultures—is a sense of authentic connection with others.

Technology addiction is increasing. This is especially true of younger employees who grew up with technology and are more likely to be early adopters. They happily use these devices to obtain instant gratification, alleviate stress, and receive personal validation. But there's a darker side to this technology use. On an episode of *60 Minutes* Tristan Harris, a former Google product manager, admitted that these devices are intentionally designed to make us addicted to them.[6] Every time we pick up our phones, we're pulling a lever in the hope of winning an exciting reward, much like using a slot machine.

Although it's tempting to think that Harris is speaking metaphorically about tech addiction, it's actually a very real thing. Every time we receive a text or status update, we get a jolt of dopamine in the pleasure system in our brains—the same system that controls addiction to drugs such as cocaine. Before smartphones existed, people spent an average of eighteen minutes a day on their computers and other electronic devices.[7] Today we're up to a whopping five hours a day,[8] during which we tap our phones an average of twenty-six hundred times.[9] About half of Americans are so addicted to their devices that they'd rather break a bone than their phones.[10]

Besides putting huge amounts of money into the pockets of device makers and technology companies, our addiction is reprogramming our minds and shaping our actions, feelings, and thoughts.[11] It's also interfering with our relationships. Author and thought leader Simon Sinek has observed that when young people experience stress, "they aren't turning to a person, they are turning to a device and social media that offer temporary relief." This coping mechanism has made us depressed, isolated, and less effective in our lives.

Two global studies conducted by Future Workplace in partnership with Randstad found that what younger workers say they want has very little to do with how they actually behave. The majority of the six thousand twenty-two- to thirty-four-year-old workers polled in more than ten countries told researchers that they prefer in-person communication to technology. Nevertheless, over a third spend approximately 30 percent of their personal and work time on Facebook.[12] Instead of having in-person meetings and phone calls, we choose texting, instant messaging, and social networking. Many of my peers even become frustrated when someone calls and leaves a voice mail, which they view as an interruption.

Workplace loneliness is spreading. When we rely on devices to connect with other humans, our relationships become weaker. Replacing human interactions with text messaging makes us lonely and unhappy. The result is an isolation epidemic that has reduced the percentage of people who say they have a close friend and has left half of all Americans feeling lonely in their public lives.[13] Dr. Vivek Murthy, a former surgeon general of the United States, told me that "loneliness and weak social connections are associated with a reduction in lifespan similar to that caused by smoking 15 cigarettes a day and even greater than that associated with obesity."[14]

To be fulfilled at work, committed to our teams, and happy, we need to focus on building deeper relationships with the people around us. The famous Harvard Grant study by George Vaillant followed 268 Harvard undergraduates for seventy-five years, collecting data on multiple aspects of their lives at different time periods.[15] Vaillant discovered that the best predictor of life satisfaction wasn't money or career success; it was strong relationships.

A few researchers have studied the correlation between the loneliness that isolated employees feel and their commitment to their teams. The consensus is that having work friends and team camaraderie can make a huge difference when it comes to job performance, loyalty to the employer, and employees' overall well-being. At Wharton School of Business Sigal Barsade interviewed 672 employees and their 114 supervisors and found that greater employee loneliness led to poorer task, team role, and relational performance.[16] In a separate study John P. Meyer and Natalie J. Allen found that the quality of employees' interpersonal relationships has a significant impact on how they perceive and connect with their companies. Employees who are lonely are more likely to feel a lack of belongingness at work and have a lower commitment to the company.[17]

Gallup interviewed more than five million people and found that just 30 percent have a best friend at work; those who do are seven times more likely to be engaged in their jobs.[18] In a separate study for this book, Virgin Pulse and Future Workplace polled more than two thousand workers in twenty countries and found that 7 percent of workers have no friends at work and more than half have five or fewer friends.[19] Those with the fewest friends felt lonely "very often" or "always" and were not engaged in their work. This is especially important for people of my generation, who consider their team their work family and their boss their work parent. No one wants to leave their family for a group of strangers at another company, just as no one wants to let their family down by being a poor performer.

Isolation in the workplace has caused employees to seek more intimacy, be more empathetic, and build deeper friendships. After surveying more than twenty-five thousand employees from ten countries, we found that remote workers who rely on collaborative tools are more likely to pick up the phone, check the tone of their emails, and befriend coworkers.[20] As an introvert who has worked from home for several years in both Boston and New York, I can relate to this need for belonging, and I know that I'm far from alone. Even with New York City's population of over 8.6 million and the countless restaurants, bars, museums, concerts, sporting events, and other activities the city has to offer, it's easy to feel lonely here. And that's a problem that affects cities and countries around the world, with devastating consequences. Japan's population is expected to drop from 127 million to 87 million by 2060.[21] The cause is fewer marriages, which stems from people not having enough human contact and instead relying on technology to do their "socializing." In France, although the average workweek is fewer than forty hours and employees receive five weeks of guaranteed vacation, the government instituted a "right to disconnect" law, allowing workers to shut down their devices once the workday has ended.[22] After discovering that more than nine million people always or often feel lonely in the United Kingdom, Prime Minister Theresa May appointed a minister for loneliness to tackle the problem.[23]

The combination of work isolation and technology overuse and addiction has given rise to what I call the *experience renaissance*, in which people are deliberately seeking out ways of spending time and doing things with others. A recent Harris Group study found that 72 percent of young workers prefer to spend more money on experiences than on material things.[24] In festivals, adult day camps, yoga retreats, group trips, and dinners, people have sought out experiences as a way to establish the connections that they crave—and miss. Despite this renaissance, the average American still spends barely thirty minutes a day in face-to-face social communication, compared to three hours watching television.[25] This lack of social connection affects not only our work experiences but also our survival. After reviewing 148 studies with 308,849 participants, Julianne Holt-Lunstad, a psychologist at Brigham Young University in Utah, found that the strongest predictor of a long, healthy life was social integration or how much we interact with people throughout the day.[26]

This book is a deeply personal one. As a young leader like you, I have struggled to maintain a balance between my business life and my personal life. I went from working on a team in a Fortune 200 company to being a "solopreneur" to being on a team again with another business, and I know that I'm guilty of overusing technology and texting instead of picking up the phone. Throughout my journey I've felt lonely, depressed, and fearful. Nevertheless, I've learned how to use technology to facilitate more in-person connections, and I know the value of those relationships and how to maximize them.

During a three-hour interview I did for a documentary on my generation, I was asked several times about the greatest challenge we face. While many might have said global warming, terrorism, or the student loan crisis, I said isolation. No question, those other issues are extremely concerning, but they're pretty much out of our immediate control when compared to the day-to-day decisions we make about our lives. My hope is to start a global movement about the importance of employee relationships and to initiate a process of making the workplace a better experience for all of us.

The point of *Back to Human* is to help you decide when and how to appropriately use technology to build better connections in your work life. I've witnessed firsthand how technology has enabled me to create a network and build a business that I never thought would have been possible. I have also seen how some of that same technology has prevented me from building deeper relationships and distracted me from living in the moment. During my interviews with dozens of prominent leaders for this book, time and time again they reaffirmed that technology is a double-edged sword. *Back to Human* addresses the hidden emotional need that makes us more human and less machine, not by discounting technology altogether but by explaining how to use it to propel your career.

My personal mission is to assist you through your entire career life cycle, from college to the C-suite. My first book, *Me 2.0*, helped you get your first job after college, and my second, *Promote Yourself*, supported you on your upward path from that first job into managerial roles. I wrote this book specifically for the next generation of leaders. I will walk you through everything you need to do to create a workplace in which your teammates feel genuinely connected and engaged. This book will help you master self-connection, promote team connection, and build organizational connection. Doing so will help you be the leader your organization desperately needs, while providing greater fulfillment for you and those you connect with.

My goal is to bring some sanity back to the workplace. We spend an average of forty-seven hours each week working, and with all our devices, it feels like we're always on the clock.[27] Because we spend so much of our lives working, it's absolutely critical that we improve our relationships with our teams and create a culture of trust.

Back to Human is designed to help you become a more effective leader by creating meaningful connections within our tech-heavy workplaces. Throughout the book you will learn how the four employee engagement factors (happiness, belonging, purpose, and trust) can be used to foster healthier and more productive work cultures. Each chapter focuses on an important topic that impacts our work lives. I start by identifying a problem and then move on to practical solutions to address that problem. You'll learn how to make better decisions about interacting with your team; how and when to use technology (and when not to); and what specific steps you can take to facilitate deeper, more effective, and more human relationships with them. The corporate cultures we're experiencing right now must change—and this book will show you exactly what you need to do to be more productive and fulfilled at work.

Cheers to your success!

Dan Schawbel

Take the Work Connectivity Index (WCI) Assessment

What It Measures, How It Works, and How to Take It

The purpose of this book is to help you build stronger relationships with your teammates so you can be a more effective leader and have a more fulfilling work experience. It's easy to get caught up in our day-to-day business challenges and ignore the important task of cultivating deeper relationships with our colleagues. We aren't self-aware about our own team connectivity because we take it for granted—yet it's essential to our success.

For that reason, I worked with Dr. Kevin Rockmann, an associate professor of management at George Mason University's School of Business, to develop the Work Connectivity Index (WCI), a self-assessment that measures the strength of your relationships at work. You and your entire team should take it to measure the level of connectivity you have with one another so you can increase that connectivity together. Teams with stronger levels of connectivity are more engaged, perform better, and are more committed to their organization's future.

Your score is based on your personal needs for social connectivity, your actual connectivity, and the strengths of your relationships at work. Upon finishing the assessment, you'll receive one of the following scores:

- **High Connectivity.** Your connectivity needs are generally being met because you're getting enough personal interaction and attention from those on your team.
- **Moderate Connectivity.** Your connectivity needs are mostly being met. Because your personal needs for social connectivity are not likely to change, you may want to keep an eye on how much social contact you're getting at work so you can continually improve your relationships.
- **Poor Connectivity.** You need more connectivity than you're getting at work. Given your needs, you likely feel isolated from your colleagues.
- **Weakest Connectivity.** Your connectivity needs are far greater than what you're getting at work, and you should make a considerable effort to improve those connections.

After completing this assessment, you will become more aware of how connected or isolated you are from those you currently work with. As a team leader who can administer this assessment to your teammates, you can identify employees who have a higher likelihood of quitting because of isolation and loneliness. You can also repeat the assessment over time to track improvements. Don't worry if you receive a poor or weakest connectivity score; over the course of *Back to Human* you'll learn many strategies that will help you improve your work relationships!

Take the assessment now at **WorkConnectivityIndex.com**.

Master Self-Connection

Chapter **1**

Focus on Fulfillment

Given how much time you'll be spending in your life making a living, loving your work is a big part of loving your life.

—MICHAEL BLOOMBERG[1]

Technology is fueling loneliness. I'm an introverted entrepreneur who sometimes spends way too long tucked away inside my home office and not enough time interacting with others. And while I've often thought that isolation and solitude give me a chance to recharge, I've also noticed that when I spend too much time alone, not only do I get lonely, but the next time I'm around people I feel somewhat awkward and stumble over my words. Those are just my symptoms. Many researchers have studied the effects of isolation on our minds, cognitive abilities, and health. Clinical psychologist Ian Robbins found that subjects who had been isolated in soundproofed rooms in a former nuclear bunker for as little as forty-eight hours suffered from anxiety and paranoia and exhibited deterioration in their overall mental functioning.[2] Social psychologist Craig Haney studied inmates who spent time isolated from other prisoners in the Security Housing Unit at the maximum-security prison Pelican Bay. Nearly all of them suffered from anxiety, nervousness, and psychological trauma.[3] Also, many studies point to social isolation and the lack of close friends as a major health risk for elderly people.

While you (hopefully) may not be able to relate to being in solitary confinement, we have all felt isolated and alone at one time or another. And it's becoming more and more common as we replace face time with FaceTime and other apps.

Technology—especially social media—is isolating us even more. A study of 1,787 young adults by psychologists at the University of Pittsburgh found that just two hours of social media use per day doubles the risk of social isolation.[4] Researchers at the University of Houston studied Facebook users, looking at how likely they are to compare themselves to others, how they feel about other people's posts, and whether they experience depressive symptoms while browsing. They found that the more active people were on Facebook, the more depressed they were.[5]

No one knows exactly why there's a connection between social media use and depression, but I have a theory. When we log into Facebook and check our friends' updates, on the surface we may applaud their achievements or be excited about their new babies, but underneath we end up feeling inadequate. That's because our own accomplishments are no longer enough. We now feel the need to surpass others and showcase our successes—one-upping others in the process—on social media. Online, we've become our best PR versions of ourselves, but I've come to believe that the more baby pictures people share, the unhappier they are. They're using the baby to cover up issues they're having in their careers or marriages. You may have friends who do this or may be guilty of doing it yourself. One recent study found that only 6 percent of young people have a completely true picture of their lives on social media, thanks to their need to impress others.[6] Although some competition is healthy, social networking has amplified our deepest insecurities about our own value. The more we check our social media feeds, the more we're comparing our lives to others. We feel that we can never measure up, and we fail to realize our own unique work contributions.

Social Media Is Hurting Our Well-Being

Social media and technology use is also associated with other negative outcomes. Gallup interviewed more than five thousand people to investigate the association between Facebook activity and real-world social activity and found that Facebook use was negatively associated with well-being.[7] Now don't get me wrong; even though I'm picking on Facebook and other social media platforms, I'm a big fan. My point is that these networks were supposed to bring us closer together, but in addition to isolating and depressing us, they have negatively impacted our well-being and have changed our view of what a meaningful career and life should look like.

This brings up a point that I'll be making throughout the book: as technology becomes more and more pervasive in our personal and work lives—and it will—interpersonal skills will become more important. "Doing business is all about relationships, and relationship-building skills will never be automated," says Dan Klamm, director of talent marketing and alumni relations at Nielsen. "Things like listening skills, empathy, conflict resolution, and follow-up will be more important than ever. Technology and social networking platforms give us new avenues to spark connections and maintain relationships, but truly building a trusting connection with someone involves 1:1 communication."

Andrew Miele, director of development at Four Seasons Hotels and Resorts, believes that this may be an especially difficult challenge for young professionals. "While technology as a medium for social interaction has been able to connect people across distances, it can lead to the opposite over the long-term. Behaviorally, generations who are brought up with technology from childhood may find it harder to engage and focus in the workplace, and may have a harder time building meaningful work relationships. Those with greater attention spans and focus, and those who demonstrate the ability to generate ideas will likely be highly sought after by future employers."

The people you regularly interact with influence your well-being, happiness, and fulfillment. When you replace emotional connections with digital ones, you lose the sensation of being present and the feeling of being alive. Every time you choose to send a message instead of picking up your phone or walking a few feet to the office next to yours, you miss an opportunity to engage with your teammates on a deeper level. Instead of letting technology be your crutch, let it be a path to more interactions, joy, and meaning.

Burnout Is Inhibiting Our Fulfillment

When we're fulfilled at work, we bring positive energy and happiness into our personal lives. We seek fulfillment through meaningful work that aligns with our values and supports the people and communities around us. That said, the workforce is currently suffering a major burnout problem. Employees are working more hours with less vacation or other time off and no additional compensation. As a result, they're changing jobs more frequently because there's less incentive to be loyal. In a research study with Kronos, we found that nearly a third of attrition is due to burnout.[8]

In another study—this one with Staples—we found that half of employees do additional work from home after their standard workday is over.[9] Managers expect their employees to answer their emails and phone calls at night, on weekends, and sometimes even when they're on vacation. Nearly half of employees don't feel that they have enough time outside of work to engage in personal activities. Unfortunately their paychecks aren't reflecting the huge increase in what is actually time devoted to the job. To make matters worse, while wages aren't even keeping up with inflation, corporate profits are up, which leaves a lot of employees feeling mistreated, unappreciated, and even more burned out.

We're also suffering from some major health issues that affect productivity and well-being and get in the way of fulfillment. One of the side effects of burnout is lack of sleep, and over a third of us get fewer than seven hours of it, whereas most of us should be getting a *minimum* of seven, according to the National Sleep Foundation.[10] We're failing on nutrition as well, with more than two-thirds of workers now overweight or obese—something that's partly attributable to eating meals alone at our desks instead of going to lunch with our coworkers[11] and partly to the increased stress burnout causes, which often leads to overeating. In fact, when we asked thousands of employees to name the biggest obstacle to work performance, half said stress.[12] It's difficult to get work done and be healthy when you're anxious, are stressed, and have a backlog of projects you need to complete.

Mental health affects our well-being and happiness as well. About 20 percent of employees suffer from mental illness, and antidepressant use has surged 400 percent over the past ten years.[13] Because employees aren't making more money, they have less to save, which makes them even more stressed out. A third of employees have trouble meeting household expenses, and of the 50 percent who carry credit card balances, one in four has trouble making minimum payments each month.[14] Finally, employees feel more isolated because they often aren't forming deep relationships with their teammates, and when they change jobs even those connections get lost, which leads to weaker organizational cultures. For this book, Future Workplace partnered with Virgin Pulse on the Global Work Connectivity Study of 2,052 employees and managers from ten countries.[15] Thirty-nine percent of our participants said they "sometimes," "very often," or "always" feel lonely at work. Not surprisingly, younger generations—those most likely to rely on technology to communicate with their colleagues—were lonelier than older generations (45 percent of Gen Z and millennials versus 36 percent of Gen X and 29 percent of baby boomers).

Better Well-Being Improves Engagement

Clearly your employees are facing some pretty significant challenges; as a leader, it's your job to do whatever you can to support them so they can focus more of their attention on getting work done and less of it on whatever it is that's stressing them out. The best way to do that is to prioritize their mental and physical well-being. As a leader, it's just as important that you focus on your *own* well-being. If you aren't healthy or happy, your employees will be affected by your condition. Unfortunately too many organizations (and their leaders) haven't focused on or prioritized mental and physical well-being.

The results of this neglect are startling. Workers with low levels of well-being are twice as likely to have high health claims costs, four times as likely to perform poorly on the job, forty-seven times as likely to exhibit high presenteeism (i.e., physically being there but not working at capacity because of illness or something else), seven times more likely to be absent, and twice as likely to have little intention to remain with their employers.[16]

What kinds of well-being programs does your company offer? Our study found that 36 percent offer flexible work hours, 24 percent offer health-risk assessments, and 24 percent offer healthy food options. Sadly, more than a quarter of employers don't offer any well-being programs at all.

The good news is that overall, workplace well-being is on the rise—in part because it makes financial sense, but also because employees are demanding change and are seeking leaders and organizations that put well-being first. It's much more common these days to find walking or standing desks, nap or quiet rooms, on-site gyms, and yoga or meditation classes in the workplace. Companies have realized that by improving their employees' well-being, they can lower health-care costs and absenteeism while increasing productivity and retention.

Employees reached similar conclusions long ago. "If there is one thing I won't sacrifice, it's my health," says Amanda Healy, senior marketing manager at TIBCO Software. "I view exercise and well-being as a luxury, and it's a luxury I refuse to live without. I actively schedule in time every day to go for a run, hit the weights, ride my bike, or attend a spin class. Without this scheduled personal time, I wouldn't stay sane (or be enjoyable to be around)."

There's no end to the ways you can incorporate wellness into your life and your company. Exercise—whether that's a seven-minute app-based workout in a hotel room, a twenty-mile training run, or the hour-long boxing classes that Kiah Erlich, senior director at Honeywell, takes—is one of the top ways people take care of their physical (and mental) health. Many make a commitment to eat healthy meals. Some meditate, whereas others, including Sam Howe, director of business development at MSLGROUP, post inspirational mantras where they can see them throughout the day. Laura Enoch, director of brand marketing and communications at Shake Shack, has a standing breakfast date with her husband every morning, and Jessica Goldberg, senior producer at Mic, keeps a gratitude journal.

What can you do to give your employees access to more opportunities for wellness? If you need ideas, they'll be glad to tell you.

Better Relationships Promote Greater Fulfillment

While conducting numerous research studies over the past few years, I have learned that as important as well-being is, being paid fairly is the top priority for all workers. In today's workplace, it's no longer taboo to talk with your colleagues about how much you make, and you can easily look up pay grades online to discover whether you're being compensated fairly. (Fairness is the trait employees value most in their leaders.[17]) Regardless of people's age, race, gender, level of education, or country of residence, money will have a major influence on what companies they work for, how long they stay there, and how well they perform. When we aren't paid fairly, we feel dissatisfied, complain, and look for new job opportunities. With so many full-time workers taking on freelance work to make ends meet (or failing to save for retirement), I don't blame them for focusing on compensation. But there's more to a fulfilled life than money.

Nobel Prize–winning psychologist Daniel Kahneman discovered that emotional well-being rises with income—but only until we make about $75,000. After that, the whole money-buys-happiness idea is largely a fantasy.[18] Although money is clearly important to workers, and you should give raises and bonuses, money alone won't help with their overall well-being. The relationships you have with others are a much better indicator of your long-term well-being. Too much focus on money—and too much staring at our phones—limits our ability to build those relationships. In fact, our devices are actually creating *weaker* relationships within our teams. Forging stronger relationships would help our well-being a lot more than money would.

Relationships with your colleagues at work are critical, not only to your employees' (and your own) wellness, but also to the long-term health of your organization. "The real power of a very strong team is that people independently understand the work that they have to do, and they go above and beyond to deliver," says Mathew Mehrotra, head of the Canadian personal banking digital experience at BMO. "I think the core of that is actually the deep connections between leaders and their teams. I have gone much farther because I am personally committed to leaders here, because I think their vision is the right vision and as people I really respect them, and I have got so much more out of my team on the same basis."

Leor Radbil, senior associate of investor relations at Bain Capital, agrees. "Having a good relationship with my immediate coworkers is extremely beneficial," he says. "First, it makes coming to work every day enjoyable. More importantly, our friendship and familiarity make working together easier. I'm comfortable asking my peers for help, and they're comfortable coming to me with questions or asking for advice." Felipe Navarro, global marketing manager at Siemens Healthineers, sums this up quite nicely: "The basis of a good performing team is trust, and trust is only developed with relationship."

If you build strong relationships with your teammates, they will work harder for you and stay longer with you, and you'll feel more fulfilled—not just as a boss, but also as a human being. Good relationships minimize bottlenecks when managing projects and make coming to work more enjoyable, even when difficult issues inevitably arise. Focusing on developing stronger relationships with your teammates will help you fulfill your own needs and help your team fulfill theirs.

Getting What We Need to Feel Fulfilled

We all have basic human needs that must be met for us to feel fulfilled. If we examine Abraham Maslow's hierarchy of needs, we see that after meeting their physiological and safety needs, people focus on belongingness and love. The relationships we have with our coworkers and friends are more important than our need for self-esteem and self-actualization. "A sense of belonging as a basic human need is greatly fulfilled in the workplace," says Nim De Swardt, the chief next-generation officer at Bacardi. "The quality of people I work with and a job that brings me true meaning is the core of my existence."

However, somewhere along the line the order got switched, and we have skipped over relationships and focused on making ourselves feel good (self-esteem) and on getting ahead in our careers (self-actualization). For example, instead of helping one of your employees with a project, you might decide to continue working on another project that you believe will help your career more. The truth is that helping your employee with their project would both strengthen your relationship and satisfy your mutual need for belongingness. This would benefit your mental state, which in turn would make you more productive and happy. In addition, your coworker would be more likely to work harder for you.

During my interview with Lazlo Bock, the former SVP of people operations at Google, I asked him why employees stay at the company. "The single biggest reason is because of the other people. They feel surrounded by curious and interesting people that want to have a big impact on the world," he said. It's not snacks, pool tables, free beer, food, or driverless cars that we need; it's people. Our relationships with our coworkers will make us stay with our companies longer and be more fulfilled. Your teammates can help you solve problems, do work that you don't have time for, and be your friends if you let them. Rajiv Kumar, president and chief medical officer at Virgin Pulse, told me that having good friends at work is essential. "Let's say something you're working on doesn't go your way, you have a bad interaction with somebody you know, or you fail at something you're trying. If you have really close friends at work who can lift you up and make that work day a more positive day, that's huge."

Successful leaders in today's working world also improve their well-being by helping their coworkers improve theirs. In a study of over five million people, Gallup found that those who have a best friend at work are seven times more likely to be engaged in their jobs, are more productive, and are more innovative. Yet less than a third of employees have a best friend at work.[19] As I mentioned previously, we're spending an ungodly amount of time working, at the expense of our personal lives.

The bottom line is that regardless of our age, gender, or ethnic background, we all share the basic human needs to connect deeply with others, to feel loved, and to matter. By meeting these needs, we will be happier and more fulfilled and thus more productive and successful in our teams. Being a leader is about creating fulfillment for yourself and your team, and when you do, the true magic at work occurs.

Pop quiz: I've used the word *fulfilled* or some variation at least a dozen times so far in this chapter. It's the kind of word that might seem easy to define but can mean very different things to different people, particularly when we're talking about fulfillment at work. What does it mean to you? Let me provide a few examples of what others say.

Derek Thompson, a senior editor at *The Atlantic*, says that his fulfillment "doesn't live in accomplishments but in the process of learning to love the process." Sam Violette, manager of e-commerce, mobile, and emerging technologies at Land O'Lakes, Inc., says that "nothing tops the satisfaction derived from positive, measurable, tangible impact on my company or co-workers." Vicki Ng, senior program manager of Global University at Adidas, needs continuous growth and learning, while Rajiv Kumar, chief medical officer of Virgin Pulse, needs an intellectual challenge. For Philip Krim, CEO of Casper, it's working with smart, engaged, empathetic people. And Rosie Perez, lead financial officer for global consumer services in business planning at American Express, feels most fulfilled when the people in her organization succeed in their careers. "My best days are when someone that works for me finishes up a big project, gives a fantastic presentation, or finds a great new role," she says.

Begin by Focusing on Your Own Fulfillment

When you're fulfilled in your job, you're moving closer to achieving your life goals. This is your ongoing journey to develop yourself and make a difference. Every time we try to compete with or compare ourselves to others, we derail our own fulfillment. If you see one of your friends sharing on Facebook that they are going to start a company, that doesn't mean you should quit your job and follow in their footsteps. They have made that decision based on what makes them feel fulfilled. What fulfills you probably isn't the same.

The cool thing about fulfillment is that it's deeply personal, and although you'll need your team's aid to accomplish your goals, at the end of the day you're the one who's responsible for your own fulfillment. There are benefits: when you feel fulfilled, you naturally have a positive attitude and a clearer direction for what you're working on.

Define Your Own Fulfillment

Answer the following questions to help you define your own fulfillment:

What do you enjoy doing the most?

What do your past accomplishments tell you about your strengths?

What are your core values (i.e., adventure, challenge, contribution, respect)?

What brings out your most positive feelings and emotions?

Where do you envision yourself in the future, and why?

Here are a few examples of how notable people define their own fulfillment:

Maya Angelou: "Success is liking yourself, liking what you do, and liking how you do it."

Richard Branson: "The more you're actively and practically engaged, the more successful you will feel."

Deepak Chopra: "Continued expansion of happiness and the progressive realization of worthy goals."

Next, Support Your Team's Fulfillment

When you're on a plane and the crew members go through the preflight safety demonstration, they always say, "If you're traveling with a child or someone who requires assistance, secure your mask first, and then assist the other person." In the same way, when it comes to well-being, once you're confident of your own needs, you can—and should—be a role model for your team. Research shows that well-being is contagious, meaning that if you have a high level of it, your positive state will begin to rub off onto your colleagues.[20]

As a leader, you're in a unique position to ensure that your employees get their needs met and to encourage them to enroll in any wellness-related programs that your company offers. But before you can do that, you need to understand what those needs actually are. And that starts with one-on-one conversations. "A lot of times the simplest way to help a teammate achieve their goals is first asking what their goals are," says Vivek Raval, head of performance management at Facebook. "It's easy to assume people's goals based on their position and your perception of their situation, but I've always been surprised at how varied the responses can be when you hear it straight from them in their voice."

Sample Conversation About Fulfillment

You: I wanted to take a minute to speak with you about your goals and what I can do to help you reach them.

Your teammate: I want to become a marketing executive at this company and be able to retire by age sixty, like my parents did.

You: That's great! Let's put together a development plan for how you can advance here and make more money, so you can retire when you want to. Let's get together every Monday so I can coach you. And I'll start giving you new projects that will help you gain more visibility and recognition here.

Your teammate: Thank you for the support. I'll send you a calendar invite so we can lock in those coaching sessions.

The Five Characteristics of Personal Fulfillment

Becoming fulfilled is neither simple nor easy. There are several key factors you need to focus on to ensure that you're living a happy, well-balanced, and meaningful life.

Let's dig a little deeper into these five areas of fulfillment.

Connection. A strong connection to your teammates makes work more meaningful and enjoyable. The lack of it makes work feel like a chore and creates the silos that eliminate creativity and innovation. You can create connection by encouraging your teammates to support one another. This might mean ensuring that you have more face-to-face conversations and joint social activities so that you can get to know each other better.

Values. If one of your values is authenticity, create and support a transparent and honest culture in your team. Be open to sharing personal information or summaries of conversations you've had with senior executives.

This will demonstrate your authenticity and help build trust. Your values are reflected in your actions so the more you demonstrate them, the more you internalize them.

Purpose. Think hard about your personal story up until right now and about the thread that connects the decisions you've made. My purpose in life is to help my generation throughout the entire career life cycle, from student to CEO. Every decision I make has to be aligned in a way that keeps me on track to achieving that purpose.

Openness. Many people fear change because it is, almost by definition, unpredictable. But as a leader you need to be open to it. When you're recruiting new teammates, look for differences more than similarities. And with your current team, being open means meeting with and including new people who have diverse backgrounds and worldviews. Instead of keeping company secrets, confide in your team so that you can build trust. Being open is also about expressing your true feelings instead of holding back. If something is bothering you, share it with your team members so they become more open to you and understand you better.

Accomplishment. This is not only something we desire but also the emotional feeling we have when something is finished. If you want to be more accomplished, set more goals and make sure they're attainable. Smaller goals can lead to bigger ones, which will give you different levels of achievement at different times.

These five characteristics have a direct impact on your work experience, health, and well-being. If you aren't fulfilled, you won't be able to help your teammates be fulfilled either. To assess how well you're doing when it comes to each of these characteristics, fill out the following report card by responding to each question with a "yes" or a "no." Then add up all the yeses.

Fulfillment Report Card	
I feel that my life lacks true meaning.	
I regularly feel bored and not challenged in my job.	
I jump from project to project without a clear sense of direction.	
I feel like I have not connected my core values to my work.	
I feel isolated and distant from my teammates.	
I rarely speak up in meetings because I'm afraid of rejection.	
I'm struggling financially, and it's affecting my work.	
I'm not confident that I can improve my current work situation.	
Total	

If you have fewer than five yeses, you have a strong sense of well-being and fulfillment. You have trusting relationships, are able to get things done, and are mentally healthy. If you have five or more, you need more trusting and deep relationships with your team, and you need to devote more thought to what gives your life meaning and pride.

Now that you understand how fulfilled you might be, you're ready to plan your future so you can enhance the five characteristics. Take a look at the chart that follows. In the "Current" column, order the characteristics according to how well you're doing. Then, in the "Priority" column, order them based on what you most need to improve on.

Characteristics of Personal Fulfillment	Current	Priority
Connection: Having supportive relationships		
Values: Aligning your work with your personal core beliefs		
Purpose: What gives your life meaning		
Openness: Your ability to adapt to people, situations, and change		
Accomplishment: Completing tasks and achieving goals		

Stop Letting Technology Derail Your Fulfillment

Tech devices trick us into thinking that they're helping us connect with others, but in reality they're a barrier that undermines and weakens our relationships. I've seen coworkers in the office text each other from three feet apart. When we do that, we miss the body language, emotion, and intensity that come from being right next to others. "It's easy to get lost and hide behind instant messages and emails and online platforms," says Virgin's Rajiv Kumar. "But true connection only happens when we pick up the phone or go meet with someone in person. That's ultimately the only way that we can be successful in our work and feel fulfilled in our work." If you just put down your phone and take a few giant steps toward someone, you'll be able to solve problems a lot more quickly.

Technology also makes it more difficult to express our values because people interpret messages and phone calls differently than they do in-person exchanges. Let's say that you value compassion. If one of your team members is having a bad day, sending an emoji may *seem* like a way to express how you feel toward your teammate, but it's not. When it comes to expressing your values and showing others how you feel, there's no substitute for basic, face-to-face, human interaction.

Technology can also get in the way of our purpose. For example, when we spend a lot of time looking at other people's social media updates, we tend to compare ourselves with them, which triggers our natural competitive instinct. Because a lot of people exaggerate their accomplishments on social media, reading about them may make us feel inadequate. That can lead us to rethink our goals and purpose (even if they don't need rethinking) or to adopt someone else's. Daniel Kim, engineering manager at Instagram, made an interesting point. "Social media makes it much easier to maintain loose connections with your co-workers," he told me. "However, it still takes intentionality and effort on both sides to build the deep, meaningful relationships that translate to friendships that last a lifetime. Overall, I'd say technology has enabled me to maintain more loose connections with my coworkers, but has not impacted the number or depth of the really meaningful work relationships that I have."

Technology can help us better adapt to changes in our environments and connect with a diverse and dispersed workforce, but it can also make us more narrow-minded by encouraging us to focus on the same type of people. We all follow our friends and our friends' friends. We join online groups of people who share our interests and goals. As a result, we're not being exposed to people with different experiences and contradictory viewpoints.

Our worldview provides the framework for our thoughts, beliefs, and actions. It's a product of everything we've been exposed to and every experience we have had, from our earliest childhood all the way through our adult lives. We're spending more and more time online consuming news, information, and ideas, which has impacted our cognitive ability, morality, and behavior. Almost as many Americans get their news online—either from news sites, apps, or social media—as from watching television (43 percent versus 50 percent).[21] On social media, we intentionally follow news outlets and personalities, but we unintentionally receive suggestions sent to us by an algorithm. We purposefully search for information that confirms our own beliefs and are fed news stories based on what social media sites want us to know. When analyzing data about the topics people discussed on Facebook from 2010 to 2014, researchers from several Italian institutions and Boston University concluded that we seek out information that reinforces our views, accept it as true, and then share it.[22] This unintentionally narrow-minded habit causes us to live in an echo chamber and reduces our ability to embrace diverse ideas and/or empathize with others. That's why we need to be aware of how technology is affecting our opinions and be open to other points of view. You can start by following a person or media outlet that you know has a different political or social view. While this may make you uncomfortable at first, by allowing yourself to open up to other opinions, you'll become more well-rounded, better educated, and open to others in and out of the workplace.

How to Discover What Makes You Feel Fulfilled

Finding your own fulfillment is a noble and epic quest. But it's important to understand that you're never going to feel fulfilled all the time and that you probably won't be able to achieve fulfillment in all areas of your life at once. Your journey toward fulfillment starts by identifying exactly where you're going. And the only way to do that is by experimenting, reflecting, and getting feedback along the way. It's taken me years to figure out what makes me fulfilled, what I'm good at, and whom I need to surround myself with so that their strengths can balance out my weaknesses.

Whenever I speak to people who are feeling lost and need direction, I always suggest that they try multiple options. The more you do, the more you realize what you don't want to do, as well as what you enjoy. When you find activities you enjoy, you can invest more of your time in them.

Whenever you try a new activity, project, or task, think about how it makes you feel inside and out; if you're glowing and feeling high while you're doing something, your brain and body are giving you a hint that you should do that more often and that it might lead to a more fulfilling career. Fulfillment is often found in benefiting your teammates. Think about how you're enriching their lives and satisfying their needs. Helping others will give you a better sense of your own purpose, path, and goals.

How to Help Your Team Discover What Makes Them Fulfilled

1. Get to know your colleagues on a personal level. The more you understand their unique situations, life goals, passions, fears, and obstacles, the more you can help them. You'll appreciate them more when you know the "real" them, and you'll have an easier time forging a long-term bond with them. Just so you know, only 20 percent of employees in the Virgin Pulse study said they were currently feeling fulfilled on the job.[23] What would make them feel more fulfilled? Thirty-one percent said more flexibility, 29 percent said meaningful work, and 26 percent said a supportive team.
2. Listen to what your employees say, without interrupting. This shows respect and will help you better solve their problems. It also demonstrates that you're willing to engage with their ideas, thoughts, and feelings. Once they have finished, summarize what they said so you can be sure that what you think you heard is what they actually meant.
3. Set boundaries with your teammates for technology use. Coworkers don't want to be added on social media unless you've already established a personal connection with them.
4. Remove obstacles that get in the way of employees' fulfillment. One often-overlooked obstacle to a team's success is having a teammate who isn't performing well or has a bad attitude. It's much better to fire employees who negatively impact team morale and prevent others from achieving their goals than to keep them around and let them poison the rest of the team.
5. Check in with your employees and ask how they're doing and whether they feel like they're on the right path. This involves

ongoing development meetings in which you mentor and guide them and give them the training and encouragement they need to be successful in their lives and careers.

6. Give your team more responsibility. This will ensure that they feel challenged, and it will help them figure out what they do—and don't—like and excel at.
7. Recognize your teammates, and see their positive qualities. People want to see themselves in a positive light, and your feedback can help guide them to a more fulfilling career.

Use Conversations to Improve Your Team's Well-Being

When your teammates feel connected, respected, and safe, they'll be more likely to stay with your company, create positive energy, and attract new teammates. You want to create a safe environment in which your teammates feel comfortable sharing aspects of who they are, what their needs are, and what their future goals are. Going a step further, it's important that you help them achieve their goals if you expect them to do the same for you. As you're doing this, consider the role that technology plays and how it has made it easier to be distant from your teammates, especially those who work remotely. "I lead a technology business, yet there are times when I can't stand technology in the workplace," says Kiah Erlich, a senior director at Honeywell. "Emails, instant messages, cell phones, conference calls—all of these are a failure point if you don't meet regularly face-to-face." In-person meetings, or at the very least videoconferences, are essential. Without them, it's easy for employees to feel insecure about what they're contributing or to feel rejected, misunderstood, or disrespected.

I have provided some sample conversations to help you better understand how to handle these critical conversations about well-being.

Conversation A

You: Tell me about your experience working on this team over the past year.

Your teammate: I feel like I'm getting paid fairly and I truly love my work, but my relationships with team members aren't strong. I feel distant from them, and it could be because I work remotely. I haven't met them personally, and it just isn't the same.

You: I suggest that you come into the office at least once a week, and we'll schedule a meeting so that you can interact with the rest of the team. We'll also do a team social event every month so that you can get to know them on a more personal level.

Conversation B

You: Our team is going through restructuring, and I'm assigning everyone a new role. The roles that are available include email marketing, social media, and mobile development. I see you as being our social media specialist.

Your teammate: I think I'm better in the email marketing role—I don't even have a Facebook account.

You: I know this change is going to be hard for you, as it is for all of us. Show me examples of some emails you've written or campaigns you've done.

Conversation C

Your teammate: I appreciate your assigning this project to me, but what's the point of it?

You: This isn't a glamorous project, but it impacts a lot of other people and gives me the chance to assess you in action. I'll get a better idea of how you work and what projects I should assign to you in the future.

The goal of these conversations is to help your employees become more fulfilled and improve their well-being. In conversation

A, you're showing your teammate that you understand their concerns and are committed to helping them improve. By asking how your teammate feels and then tackling the issue at hand and solving it, you're showing respect and caring. In conversation B, you focus on the important skill of adapting to change and how it leads to personal fulfillment. You'd be surprised at how often organizations restructure and roles change. With three new positions available, you want to ensure that you're filling each one with the most appropriate person. Although you may have thought that your teammate would have been a perfect fit for the social media role, they clearly think differently. By asking for evidence that they have the right skills, you're giving that teammate a chance to prove that they could be successful in that role. In conversation C, your explanation of why the project is important will help your teammate better understand its significance and their role in making it a success. In all three conversations, you have to listen to your teammate's needs and then figure out how to satisfy them if possible. And sometimes you won't know what those needs are unless you directly ask.

Embrace Work-Life Integration

To be more fulfilled and to develop a deeper sense of your own well-being (and your team's), you need to shift your mind-set. Back when I had a full-time job at a Fortune 200 company, most employees were working more than forty hours a week. When I interviewed the now former head of human resources, he said something that I'll always remember: "The line is blurry between work and life. So we have to make sure that our people can do personal stuff while on the job, because I know they're doing work things when they're off the job." After he said that, my entire view of work-life balance changed. A decade later I sat down with Richard Branson, who reinforced this new reality. "There should be no difference between somebody's life at home and somebody's life at work. If at home, you feel that the environment is important, it should be important in the workplace. If at home, you've got friends, you should have an equal number of friends in the workplace."[24]

The fact is that work-life balance is a myth, in large part because the phrase itself implies that there are only two components to our lives, work and life; that the two are separate; and that we'll divide our time equally between them. "Equal balance is imperfect, and I believe that the more we try to create the balance, the more stress we add to ourselves, which in turn could create another unbalanced situation over time," says Justin Orkin, head of advertiser demand platforms for the western United States at Oath, a Verizon company. The days of coming home from the office and unplugging are over. We now live in a 24/7, always-connected business environment. Your company doesn't stop running when you leave the office or when you go on vacation.

That's why, instead of work-life balance, I prefer to think of work-life *integration*, which creates more synergies among all areas of our lives and puts us in control of how we allocate our time. In other words, it's the ability to blend what you need to do personally and professionally to make both areas successful for you. Clinical psychologist Maria Sirois says that work-life integration results in less stress and more fulfillment.[25] When I spoke to Denise Morrison, the CEO of the Campbell Soup Company, she admitted that achieving a perfect balance isn't always possible. "I've always approached it as work-life integration. You can do it all, just not all at once. It's possible to integrate priorities at home with the priorities of your career."

As a leader, you need to support those employees who need to take personal time while at work, who have to call their aging parents, or who need to take the morning off to be with their children. On the flip side, they need to commit to getting the work done, even if that's later at night, earlier in the morning, or on weekends. Work-life integration requires trade-offs. "I firmly believe that if you want to progress, grow, and hold senior positions with a lot of responsibility, you will never have true work and life separation," says Ilona Jurkiewicz, vice president of talent and development in early careers at Thomson Reuters. "Instead, I view both as a symbiotic relationship. If I need to take care of something personal in the middle of the day, I do it. If I need to work late a few nights, then I hold off on personal needs and I do that."

Three Secrets to Greater Work-Life Integration

1. **Respect your boundaries.** Know your boundaries and communicate them to your team members, while getting to know—and accepting—theirs. If you need space during lunch to reflect on your life, or you need to take your child to school every morning, then speak openly about it with your team. At the same time, ask them what their boundaries are so you know what topics are off limits, what their personal needs are, or what they might need to attend to during the workday. "Boundaries are less clear, as my phone comes with me everywhere, and I am always reachable," says Jennifer Fleiss, cofounder and head of business development at Rent the Runway. "On the one hand, this enables increased efficiencies and abilities to work from home or leave the office and get work done later. On the other hand, I will need to make a conscious effort to focus on my kids and not always be distracted!" For Stephanie Bixler, vice president of technology at Scholastic, boundaries are very clear. "My goal was to spend at least two hours with my daughter every day—30 minutes before work, and 1.5 hours after I got home and before her bedtime," she told me. "While I wanted to spend time with my daughter, it was not just about 'want' but about 'need,' as my childcare only allowed for coverage during business hours."
2. **Control your devices.** Take control of technology by turning off your phone at certain points in the day when you need a break from work. In our society, turning off your phone is a sure sign that you're off duty. When it's on, *you're* on. This won't be easy, but start by trying to turn off your devices when you're having lunch with your colleagues or when you get home from work. At some point you may want to make a clean break between work and personal. When Adidas's Vicki Ng is on vacation, she

actually deletes the email app from her phone and asks people to text her if there's something urgent. Personally, I'm not sure I'm brave enough to do that!

3. **Own your schedule.** Find a schedule that works for you and allows you to be fulfilled. If you get your best work done in the morning, get to the office earlier and leave earlier. If one of your coworkers feels the same way, give them the same flexible-schedule option. Your team members are charged with getting results for your company, and you should support whatever helps them accomplish that goal.

Key Takeaways from Focus on Fulfillment

1. **Focus on relationships over achievement.** The stronger your relationship with your team, the more you'll be able to achieve. When it comes to our life needs, strong and deep relationships with our coworkers, with whom we spend at least forty hours a week, are much more important than what we accomplish. If we have weak ties, it's more difficult to get anything done or to achieve our—and their—goals.
2. **Identify what makes you feel fulfilled at work and have one-on-one conversations with your team members to learn what makes *them* feel fulfilled.** You need to be aligned with your team so you can help them achieve their goals, not just yours. These discussions on fulfillment will make people want to work for you longer because you'll have proven that you're committed to their success.
3. **Strive for work-life *integration* instead of balance.** Think about the most important activities you want to accomplish in your work and personal life, then structure your day around them.

Chapter 2

Optimize Your Productivity

If you put out an ordinary effort, you're going to get ordinary results. If you put out an extraordinary effort, you're going to get extraordinary results.

—Steve Harvey[1]

We have been equipped with an assortment of devices and apps that on the surface make us feel more productive and like expert multitaskers. Yet underneath, despite all the positive outcomes that these devices have promised, they have distracted us, drained our productivity, and consumed every last bit of creativity from our minds. In the past ten years companies have become slower to hire and have consolidated their teams, yet the pressure to innovate and compete has accelerated. The slogan "Do more with less" has become the driving philosophy of companies around the world. The pressure on employees to become simultaneously more productive and more efficient has now become counterproductive. Employees are burned out, unhappy, and tired of working longer and harder for no additional pay and less personal time. As a result, they're switching jobs more quickly. In one study we found that nearly half of all attrition is due to burnout, and each loss costs employers thousands of dollars in health-care-related expenses, productivity losses, hiring, and training to fill vacancies.[2] While using connected devices to maximize our resources may seem like a good idea, the consequences of doing so have been dismal.

Today, text messages and emails are competing for our time with human connections, and the technology is winning big-time. Several studies show that the average office worker receives more than one hundred emails per day. We used to complain that phone calls and random office visits were distracting, but those are nothing compared to the ridiculous number of messages we receive through our devices. The more messages we get, the longer it takes to review and respond to them. Unfortunately there are still only twenty-four hours in a day. The only thing we can control is how we spend our time.

The overstimulation we receive from using these tools has made us underutilize our cognitive abilities, which are critical for optimizing our own productivity. When you constantly hear (and feel) alerts from your phone, you get aroused and excited, which takes your mind off your work and focuses it on something else that may be less important. Professor Gloria Mark, at the University of California, Irvine, says that these notifications are hurting our ability to concentrate on individual tasks. Since 2004 she has followed employees around with stopwatches and timed their actions. In the early days of her study, she found that they switched their attention every three minutes. In 2012 it was just over one minute, and in 2014 it was less than a minute.[3] We've become slaves to the very devices that were created to be obedient to us, and that has made us lose focus on what's important in and out of the workplace.

Even though these tools make it easy to manage our calendars, track our tasks, and quickly message a colleague, we're so barraged by notifications that we lose track of time, and before we know it, the workday is over. Nearly 100 percent of employees admit to being distracted during the workday,[4] and nearly 60 percent of these interruptions involve some sort of messaging device.[5] This should matter to you both on a personal level and as a leader, because when they're all added up, those interruptions cost companies an average of more than $10 million annually, or just over $10,000 for each employee. By removing distractions and staying focused on one task at a time, we can be less stressed, more productive, and happier at work.

As a leader, you should encourage your employees to spend more time engaging with one another in person and less time communicating via technology. Researchers Mahdi Roghanizad and Vanessa Bohns had forty-five participants each ask ten strangers to complete a survey.[6] Every participant had to make the same request using a consistent script, but half would do so over email and the other half would do it face-to-face. Before the participants started making their requests, they had to predict how many of the strangers would agree to finish the survey. Both groups estimated that about half of the ten people they approached would agree to take the survey. Turns out that everyone was wrong. So how did it turn out? Face-to-face requests were thirty-four times more successful than email requests in getting people to fill out the survey.

Danny Gaynor, former chief speechwriter at the US Agency for International Development, gave me a great example of the power of face-to-face communication over email. The day before a major announcement (the launch of the largest-ever global coalition of nations committing to ending premature death for kids under five years old), Danny and his boss spent a huge amount of time working on the boss's presentation, texting, sending emails, and exchanging PowerPoint decks. Dozens of leaders, from Kenya to India, were flying in to attend the unveiling of this critical initiative. Danny felt isolated and stressed out while creating a presentation laying out the US vision to save millions of children's lives around the world. (Can you blame him?)

"I'll never forget standing backstage, with prime ministers and cabinet officials waiting just feet away, with my boss. Because we hadn't been in the same room—because we had been working via technology—we had never run through his presentation until a few moments before he would go on stage. I frantically deleted photos, moved graphs, and rewrote sensitive words about Afghanistan and Nepal and Colombia. My boss chugged a glass of water, looked at me, took a deep breath, and said, 'Good job on the speech,' as he walked into the stares of the world's most powerful people."

This experience taught Danny about the importance of face-to-face communication in addressing critical matters, which has made him more effective in his current role on the Narrative, Innovation, and Executives team at Nike. "The feedback I received was slow, confusing, and uneven," Danny told me. "Had I been able to get in a room with the boss—face-to-face, even for 10 minutes—I could have accomplished what took hours through technology platforms." Technology can help—but sometimes, for those big projects, there's nothing more powerful than a few people, hovering over a laptop, figuring out just the right way to explain something important.

It's ironic that even though email is the most common way we communicate at work, in many cases it's also the most counterproductive, and it routinely gets in the way of face-to-face interactions that lead to greater productivity. "Email is the number one enemy of productivity," says Stephanie Bixler, vice president of technology at Scholastic. "People rely too heavily on email as a means for escalating issues without seeing it through to the solution, and it obfuscates responsibility."

In my company's research, over half of employees say that more live conversations would reduce the number of emails they receive.[7] And in the Virgin Pulse study, almost a third said that spending more face time with coworkers would make them more productive.[8] "A lot of context may get lost [over] email," says Ulrich Kadow, CEO and chief agent of AGCS Canada at Allianz. "Picking up the phone or speaking with colleagues in person is often the best and quickest way to resolve conflicts." Katie Vachon, merchandise manager of women's apparel at Puma, agrees. "We can sit in the same office and send emails from our respective desks without just walking over and talking to people," she says. "This leads to even more emails being sent and causes confusion." So instead of going back and forth on email and praying that the idea you're trying to convey actually gets received and produces the effect you want, meet in person and take a few minutes to explain what you want and why it's important.

Managers have to cope with having fewer resources, yet more pressure, to accomplish bigger goals. A study by Bain & Company estimates that a typical manager, over the course of a forty-seven-hour workweek, has fewer than seven hours of uninterrupted time.[9] A full twenty-one hours of her time goes to meetings and another eleven to managing email. Managers don't have time to get their own work done, much less to think creatively. About thirty years ago the average manager received one thousand notes in a year about having missed a phone call. When voicemail became a thing, that manager might have had to listen to about four thousand messages in a year. But today we're talking about thirty thousand communications in a year, thanks to the variety of ways we can receive them (IM, Skype, FaceTime, email, text, voice mail, etc.) and the number of devices we rely on.

"Technology can be a powerful distraction that pulls managers' energy away from their people and into their screens," says Paul Reich, SVP of local sales at Yelp. "Front-line managers in particular are susceptible to being sucked into their screen for hours on end, when the real work is out there on the floor, interacting in person with their people. We try to teach our managers that there's no more powerful computer than your mind, no better listening device than your ears and speaker than your mouth, so clamshell that laptop and start observing and commenting in real life."

Use Technology to Increase Your Productivity

Relying too much on technology inhibits our ability to connect and get things done. That said, used appropriately, technology can be our greatest ally. I've found that there are several key ways to use technology to bring us together so we can get our work done in the quickest, most efficient way possible. In many cases technology can be used to promote face time among your team members.

1. Use a conference room booking system to lock in a time and place for your team to talk about an important project or facilitate a one-on-one catch-up meeting.

2. Use a calendar application to schedule appointments and get everyone on the same page to meet at a single place at an agreed-upon time.
3. Use search engines to quickly answer basic questions. That will save you and your team from having long, unnecessary discussions about them.
4. Use your out-of-office email autoresponder when you're on vacation or on a work trip so that you don't have to be on call to respond to every email when you're away.
5. Use a calendar to block off time for exercise, meals, breaks, one-on-one meetings, and your current projects.
6. Use collaborative applications as a virtual "watercooler" to quickly and efficiently generate ideas for a project you're working on or for feedback on how to improve.
7. Use a shared calendar to keep track of when employees are at the office, on a trip, on vacation, or busy so you can find the best way to communicate with them and schedule meetings.
8. Use a project management tool to keep track of team milestones and goals to stay on the right track—something that can often be unmanageable without technology.
9. Use videoconferencing to connect remote employees so they get to see one another even though they aren't physically with the rest of the team.
10. Use a to-do list, or a simple goal sheet, to keep track of what you need to do and when you need to do it by.

A problem many of us have is that we overuse and abuse the technology we believe is helping us be more productive. For instance, it makes perfect sense to text your team to register for a meeting, but if during that meeting you continue to text people who are right there in front of you, you really need to stop.

Although many blame my generation for being early adopters of new technology, we're all guilty of abusing it. We convince many older workers to use these tech tools because that makes it easier for us to communicate with them. While it may seem that using technology makes it easier to manage projects, files, and communication, our attention spans have decreased, and the nonstop barrage of digital distractions has squashed our productivity. This chapter will help you understand what causes these distractions and how to best optimize your time so that you can become more productive and happy at work.

Self-Assessment: How Digitally Distracted Am I?

I hate to break it to you, but there's a good chance that you're addicted to your devices. And like most addicts, you may not even be aware of it. A lot of us think that technology is enhancing our productivity, but we're completely unaware of the hours we're wasting reading random news websites and messaging our friends on social media. This brief quiz will give you an idea of how distracted you are. Put a check mark next to each statement that rings true for you.

Statement	
1. I turn to a device first when reviewing a new project with a team member.	
2. I manage all my projects using a device.	
3. I think in-person meetings are a complete waste of my time.	
4. I avoid making phone calls and prefer to text instead.	
5. I spend more time looking at my phone during meetings than contributing to the discussion.	
6. I find myself waiting for the next phone notification.	
7. I'm an expert multitasker who uses multiple devices at once to accomplish work.	
8. I would rather give a webinar or other virtual presentation than give one in real life.	
9. I have multiple windows or apps open when I'm doing work.	
10. I sometimes view my coworkers as distractions instead of assets.	
Total__________	

After you've finished taking the assessment, add up all the check marks. If you have seven or more, you may be underperforming at work. If you score below five, you're more in control of the tech you're currently using, and it hasn't negatively affected your productivity. In fact, it might even be helping it!

Three Actions to Optimize Productivity

Technology undermines our productivity in significant—and insidious—ways. Following are three ideas on how to curb that problem.

Action One: Procrastinate Less

When we want to avoid unpleasant tasks or at least delay the inevitable, we often turn to our phone and play a game or read a news article. When we lack structure, are anxious, or feel unsure of ourselves, we often change tasks from whatever is stressing us out to something more entertaining. It's easy to fall into the procrastination trap, and it's getting easier by the day, with an amazing number of apps and websites that we can turn to any time we'd like to disappear from work and take ourselves to a new reality. By procrastinating, we're not just losing our productive time. We're also losing our personal time because, after all, that's when we need to make up the work we should have done during work hours. The absence of personal time quickly turns into unhappiness and burnout.

Activity: Eliminating Procrastination

Whenever you're working on a project, it's always helpful to break it down into smaller tasks. When you do that, the overall project becomes less daunting and more manageable, which makes it harder to procrastinate. Let's say that you have to do a feasibility study on a new service that your company might offer. Here's how the breaking-down process might go:

1. Ask your manager about the scope of the service and any other pertinent information that will aid your research. This will

ensure that the two of you are on the same page, which will help you meet or (hopefully) exceed expectations.

2. Create a timeline showing when you need to finish the project, along with a number of achievable milestones along the way. Put the milestones and the final due date in your calendar and set alerts.
3. Identify competitive offerings. Knowing what your competitors are selling will help you assess the size of the market, allow you to identify ways to differentiate your service, and show your management that you're up-to-date on what's already out there.
4. Collect research from a variety of industry and other resources and put it in a master file so you can incorporate it into a presentation.
5. Put the research, along with your ideas, recommendations, charts, and illustrations, into a presentation file.
6. Call a team meeting and make the presentation. Ask for feedback.

Action Two: Resist Perfectionism

Technology has given us a nearly unlimited ability to edit, adjust, remove imperfections, and generally make everything perfect. The problem is that the more time we spend editing photos and ensuring that our status updates make us look successful and happy, the more time we waste and the less we get done. I'm sure you'll admit that at least some of the time you've spent changing photo filters or revising a one-sentence status update to your team could have been better spent on more important tasks. What those of us who struggle with perfectionism overlook is that it's our flaws that stimulate our creativity, make us unique, and create stronger bonds.

The fact is that there's no such thing as perfection, so demanding it from yourself or others is a waste of time and energy. I believe that perfectionism is a weakness disguised as a strength. We think that being perfect will enable us to be more productive and successful at work, yet striving for it depletes our time, causes anxiety, and makes us unhappy. Perfectionism doesn't work in the fast-paced, always-on world that we live in. If you move slowly at work, another worker who's willing to work smarter and faster will replace you. Perfectionists may take hours to send one email and end up working longer hours to accomplish the same thing as nonperfectionists. How do you think that affects your company and its ability to compete with other organizations that have fewer perfectionists?

Our quest for perfection includes finding the perfect answer to all our questions—and while technology can help with this, there is, as usual, a downside. "The Internet is a machine. Push a question in, and it spits an answer out. But what happens when the answer you're looking for isn't online?" asks Kiah Erlich, a senior director at Honeywell. "The 'no answer found dilemma' results in the following stages: 1) Confusion: Surely I didn't type the right keywords; 2) Frustration: Why am I not getting what I need?! 3) Panic: Oh no! The answer cannot be found! and 4) Realization: Humans were given brains before computers were given search engines. This over-reliance on technology and the expectation that someone else has already solved our problems has not only drained society of creativity and critical problem-solving skills, but has also handicapped us in the most critical skill needed to be a successful leader: social interaction."

When you're a perfectionist, you get stuck in an endless loop of trying to edit something when you just need to let it go. Your failure to deliver hurts you, your team, the company, and your customers. To prevent this, set consequences for yourself and anyone else whose constant revising might inadvertently delay the delivery of a project. For instance, you won't be allowed to telecommute for a month, or you warn everyone that delays in delivering future projects will impact their bonuses.

When You Have a Perfectionist as an Employee

Leader: Would you please design a graphic for the new advertising campaign we're launching in April?

Employee: Yes, I'll have a draft soon for your review.

Two weeks later...

Leader: I'd like to see a draft of the graphic you promised me two weeks ago. We need to have something finalized soon so that we can get final approval from senior management.

Employee: I'm still deciding what font to use, so I'm going to need more time. How about I give you something in one week?

While this may sound completely ridiculous, it happens all the time in the workplace. If picking a font was such a big deal (and chances are it wasn't), the employee could have saved everyone a lot of anxiety by running a few options by the manager a day or two after getting the assignment.

Action Three: Stop Multitasking

We may try to convince ourselves that we're multitasking when we're checking email on our computers, doing a status update on our phones, and participating in a conference call at the same time, but we're not. You can't really pay attention to what someone is saying when your brain is focused on something else. This is especially true when technology is involved. Are you honestly fully there in a meeting if you're texting your friend at the same time? That superhero-like ability to do multiple things at once is a figment of your imagination. Dozens of neuroscience research studies prove that our brains don't do tasks simultaneously. Instead, we jump from task to task rapidly. When we shift from the conference call to the status update to the email, there's a stop-and-start process in our brains that causes a momentary lag between steps when nothing happens at all. You have to actually stop speaking (or slow waaaay down) to send that text. You might not notice the lag, but it's there. Bottom line, instead of doing three tasks at once to the full extent of your ability, you're doing three tasks poorly.

How to Prioritize Instead of Multitask

The best way to keep from wasting a bunch of time trying to multitask is to become an expert at prioritizing your workload so you're always focused on the right project at the right time instead of on multiple projects all at once. Start by not saying yes to everything people ask of you, because there is truly no way to manage everything at the same time. Instead, focus on what's most important and delegate the rest to your teammates.

Derek Baltuskonis, director of talent acquisition at Intuit, learned the art of prioritization with the help of his manager. "My manager really helped me understand how to work smarter and more efficiently by looking at all the work I had and learning how to sequence the work and prioritize what is most important for the business and do that really well," he told me. "Whenever something new comes up, I try to think about it as not just the level of importance of the work, but to balance that against what won't be able to get done and how critical that is." Of course talking about prioritization makes a whole lot of sense, but in practice it can be complicated by competing demands. The way Derek handles these situations is to get a project 80 percent complete and then get feedback before moving forward. Seeking 100 percent perfection slows him down.

What Really Impacts Our Productivity?

Although some still believe that technology is supercharging our productivity or that multitasking is a reality, the truth is that our coworkers are the key ingredients. Coworkers who are experts in their subject matter, intelligent, kind, conscientious, and strong in work ethic will help you become smarter and more productive. When we asked employees what, besides salary, motivates them at work, more than half said, "my coworkers." And when we asked what inspires them to be creative at work, more than 40 percent said, "the people I surround myself with."[10] Keep in mind, though, that this works the other way around as well. If your coworkers are lazy, dumb, and annoying, you'll be less productive, less efficient, and less likely to get interesting projects, increased responsibility, or promotions.

Knowing all this, it's obvious that hiring the right employees will help increase your entire team's output, creativity, and productivity. Look for candidates who get along with your teammates and have demonstrated consistent progress in their careers (which is a sign of a strong work ethic and ability to focus). Besides asking about their skill sets, ask about their preferences for work environment and the types of teammates who bring out their best work. Then set up one-on-one meetings with everyone on the team to see how they get along.

Generally, the most productive people are conscious and deliberate about how they spend their time and build structure into their day, every day. Charles Duhigg, author of *Smarter Faster Better*, told me "the people who are most productive, are people who think more deeply than other people, about what they are doing and why they are doing it."[11]

Being Productive While Telecommuting

The image a lot of people have of a telecommuter is of a couch potato like Homer Simpson, drinking a beer, watching TV, and petting the dog while he's supposed to be working. The truth, however, is a very different story. Researcher Nicholas Bloom did a study of call center workers at a large travel website. With the approval of the company's CEO, workers were given a choice of working at home for nine months or staying in the office. While the company expected to see productivity decline, the exact opposite happened. Remote employees took 13.5 percent more calls than the office-based staff. As surprising as that sounds, it came as no surprise to the employees. In our global research with Polycom of over twenty-five thousand employees, more than 60 percent of those who worked remotely said that their work arrangements increase their productivity. What accounts for that increased productivity? According to Bloom, there are quite a few explanations. He attributes a third of the increase to the employees having a quieter environment. The other two-thirds he attributes to the employees putting in longer hours. "They started earlier, took shorter breaks, and worked until the end of the day," he wrote in the *Harvard Business Review*.[12] In addition to all that, employees used a lot less sick time and quit their jobs at half the rate of those who worked in the office.

But while remote workers are definitely more satisfied (despite stereotypes to the contrary), they're also more isolated. As a result, they seek out other ways to stay connected with their fellow humans. Thirty-five percent of remote workers said that they check in with their colleagues more often, and 46 percent said they pick up the phone more often. (That includes 38 percent who said they use email less and the phone more.[13])

How common is remote work? Well, the data indicate that there's a good chance that you've worked remotely at least once in your career. Nearly 75 percent of workers say that their company offers telecommuting, and about a third regularly work remotely.

If You're Working Remotely	If You're Managing Remote Workers
1. Eliminate all potential distractions by working in a room where you don't have a TV or any unnecessary technology. 2. Dress like you're at the office. That sounds silly, but it'll help you feel more professional and get you into the right mind-set. 3. Follow a regular routine so that you get in the habit of waking up, doing work, and taking a reasonable number of breaks. 4. Create a daily to-do list on which you can check off goals as you complete them.	1. Set proper expectations about their job responsibilities, due dates for projects, and how best to communicate updates on their work. 2. Require at least one team meeting each week so you can make sure that everyone's on the same page and keeping up with their work. 3. Use videoconferencing to create more meaningful interactions with them and to encourage them to dress and act professionally.

(continued)

If You're Working Remotely	If You're Managing Remote Workers
5. Remove clutter from your workspace so you can focus on your work without distractions. 6. Set boundaries. One of the reasons people who work remotely are more productive is that they never go home (in large part because they're *already* home). As a result, they tend to put in longer hours and check their work email at 2:00 a.m. on the way to the bathroom.	4. Encourage at least one face-to-face meeting each month so that you, the remote workers, and the rest of the team can build stronger relationships with one another.

The Art of Optimizing Productivity

Everyone's mind, body, and habits are different; so is what motivates them to be productive. For instance, I'm more driven by making an impact on the world, building my brand, and helping others than by making money and funding a retirement plan (although I try to do both). Younger workers tend to value flexibility and meaningful work, whereas older workers are more focused on saving for retirement and health-care benefits. Women care more about maternity leave than men do (for obvious reasons), although paternity leave is quickly becoming a staple in many benefit programs.

Because our brains are unique, so are the circumstances under which we're most creative or do our best work. You may perform better in a corporate office, whereas my best ideas come when I'm telecommuting from home because I'm less restricted that way.

Productivity outcomes also differ depending on the nature of work. How you operate as a leader in the marketing organization of a ten-thousand-person company in Asia will be very different from how you perform in the same position at a smaller company in the United States. And whereas some people are more productive early in the morning, others may not kick into action until the late afternoon.

Such differences notwithstanding, the following chart summarizes research by my company and others that will give you a general sense of how you can maximize your—and your team's—productivity.

The time each day we're most productive	Between 10:00 a.m. and noon[14]	We're more productive earlier in the day, before lunch.
The day of the week we're most productive	Tuesday[15]	Monday is when we catch up with email and tasks from the previous week, while we can begin to focus on this week's work on Tuesday.
The optimal amount of sleep for maximum productivity	Between 7 and 9 hours[16]	Sleep helps our mood, makes us more attentive, and gives us more energy.
The optimal number of work breaks	One every 52 minutes[17]	Our attention spans are short, and we can focus and do our best work for just under an hour at a time.
The optimal length of a break	17 minutes	It doesn't take long to rejuvenate, and breaks are key for resting our brains.
The optimal amount of exercise	At least 150 minutes a week[18]	By exercising, we alleviate stress, get in better shape, and reduce both burnout and health issues.
The optimal number of calories	2,700 a day for men and 2,200 for women[19]	To maintain our current weight or lose a few pounds, we should eat healthier by eliminating processed food, sugar, and grains from our diets.

There are factors that can either increase your productivity or decrease it. When you aren't physically or mentally healthy, for example, you're going to have trouble concentrating on your work because you'll be preoccupied with thinking about your health. However, when you're working with the right team in the right environment, everything seems to work better. I worked with Staples Business Advantage to study the factors that most influence productivity. Here's what we found:

Increases productivity	Collaborative environment, break time for employees to refresh, and more flexible schedules
Decreases productivity	Physical or mental illness, burnout, poor technology, office politics, limited IT support, and too many meetings

Forming New Productivity Habits

When it comes to increasing productivity, a lot of us are drawn to one (or two or twelve) of the apps that promise amazing results. There's no question that technology is capable of improving our productivity, but I've found that quite often, tech tools that are supposed to make my life easier end up taking up even more time. My goal is to simplify my life, so I focus on the tasks that yield the best results and use the fewest tools possible. The more tools you use, the more complex your daily life will be, and that complexity will keep you from accomplishing your goals.

So instead of relying on the latest tech gadgets to try to increase your productivity, the best approach is to create some new productivity-optimizing habits. According to professors at Duke University, habits account for about 40 percent of our behavior each day, so it's important to form ones that lead to the most productivity for you and your team.[20] Over the years I have tested a variety of strategies and habits, trying to figure out what works best for me and what could work best for other high performers. I discovered that for me, it's all about the morning routine. I get up at 7:30 a.m., cook breakfast, review my goals for the day (which I set either at the beginning of the week or the night before), go for a three-mile run, shower, and then start doing the work that requires the most intellectual capital. Knowing that I'm going to cook a solid breakfast that consists of an egg white omelet and fruit motivates me to wake up earlier. Setting goals keeps me focused, while the run keeps my energy flowing. Of course my routine can—and often does—shift, depending on meetings or conference calls that I can't schedule for any other time, but it's fairly consistent each day, which helps me maximize my time by getting the majority of my work done when I'm at my very best.

Aligning Your Habits to Your Goals

Ideally, your habits and your goals should align. That means that if you want to lose weight, you'll need to make healthy eating and exercise regular habits. Because habits are fairly useless without goals, let's spend a few minutes talking about goals.

Over the past five years, I have come up with my own goal-setting system. Before I had this system, I didn't have any structure and was bouncing around, doing whichever project made the most sense at the time. Today I use a Microsoft Word document to handle all my goals. That may not sound fancy, but it works well for my needs. I divide the document into three sections: daily goals, annual goals, and future goals.

Even though it's the middle category, I start with my annual goals, and I write in five professional and five personal goals for the year—no more, no less. Putting a graphic check box next to each item somehow makes it more likely that you'll want to accomplish the goal. These goals should be achievable based on what you've previously accomplished, yet challenging enough that you'll improve yourself, your relationships, and your career. Make sure you can measure the outcome of each goal. For instance, instead of saying, "I'll write an article," put a number to it and say, "I'll write 20 articles." That way, you can mark down when you write each one until you hit twenty. After you complete your annual goals, think about what you need to do today to get yourself closer to achieving those goals. For me to achieve my goal of completing this manuscript, I needed to write all the chapters, so my daily goals often included something like "write five pages of chapter 3." If I want to keep moving toward my goal of visiting a new country this year, another one of my daily goals needs to be "research travel destinations." Use your daily and annual goals as starting points for your longer-term life goals, which will go into the future goals column. Any time you think of a lofty goal and you don't have time for it now, put it in future goals. Having that list of future goals will inspire you to achieve more in your career and life.

Goal Sheet Example		
Daily Goals	**Annual Goals**	**Future Goals**
• Research for chapter 1 • Create survey questionnaire for a new study • Book flights for Greece	*Professional goals* • Finish book manuscript • Conduct 6 research studies • Speak at 10 conferences • Write 20 byline articles • Curate 4 executive events *Personal goals* • Travel to 1 new country • Volunteer at 1 nonprofit • See 1 Broadway show • Take 2 cooking classes • Make 5 new friends	• Start a nonprofit • Write another book • Create a podcast • Film a documentary

Just because my goal sheet works for me doesn't mean that it will suit your needs. Amanda Fraga, vice president of strategy and insights at Live Nation, has a more simplistic approach that ensures she stays active and invests her time in the things that are most important to her in a given month. Here are a few examples of her goals:

- One book per week (routine: listen to audiobooks on way into work)
- At least one yoga class a week (shoot for Saturday morning; if that's not possible, plan for another day that week)
- Cook dinner at least three times per week
- One date with my boyfriend at least once a week
- One learning adventure every week (networking dinners, visiting a museum, etc.)
- One monthly dinner club with friends

Create a list of goals that work best for you in your situation. Once you have your goals set up, take some time to think about what habits you may need to create or hone to achieve each of those goals. But don't go crazy with this. It's hard to work on more than one habit at a time. For each new habit, follow the process outlined below. Once you've completely mastered it, move on to the next one.

Creating New Habits

Creating new habits is a three-step process:

1. **Start small.** Instead of running for one hour in the morning, run for twenty or thirty minutes. Or instead of focusing on your biggest work project for two hours, start with fifteen minutes. When you start small, achieving your goals is more manageable and less intimidating, which makes it more likely that you'll commit to it. Pick one habit at a time so as not to overload yourself, and

make sure that it's something you can motivate yourself to do and achieve.

2. **Expand the habit.** Now take that half-hour run and double it, or double the fifteen-minute focus on a work project. After a week or two, it'll be easier to expand your habit to be more challenging and more fulfilling.
3. **Group similar habits.** If you already get coffee every morning from a local shop, build that into your exercise routine by running or jogging there and back. If you make time to meet one of your direct reports every day, schedule a breakfast or a walking meeting, which will be good for your health and will help strengthen your connection with your teammates.

Staying Creative While Being Productive

On a typical day we receive countless notifications from our devices. Each comes with a distracting beep, ding, or buzz, regardless of whether it's a text from a teammate who needs help or from Mom saying that she loves you. We are so inundated with these notifications—and so addicted to receiving them—that when we don't, we think that either there's a technology glitch or no one loves us. While some of these notifications may be useful (e.g., if there's a family emergency), others are downright frustrating, such as Johnny "liking" your social media picture. By the time you review and filter the notifications, you've wasted minutes or even hours. A study by Deloitte found that people look at their phones forty-seven times a day on average; for young people, it's more like eighty-two.[21]

Spending all that time checking our devices, in addition to the demands of the workplace, prevents us from freeing up capacity in our brains to think creatively. Our brains are constantly processing information, and we need more time to use our imagination and harness our creativity. But you prevent yourself from having downtime when you're too busy getting updates and responding to messages.

How to Be More Creative

1. **Turn off notifications on your devices.** If there's an emergency and a friend or family member needs to get a hold of you, they'll call you or stop by your office. By turning notifications off in your device settings, you can free up your mind to think creatively and focus on the projects that matter most.
2. **Embrace diverse ideas.** Although I discuss this in depth in chapter 4, here I want to point out that surrounding yourself with people who have different perspectives and worldviews will challenge you to open your mind in new ways. Go to events outside your industry and to places that are unfamiliar to you, such as museums, the theater, or a class on a subject you've always wanted to learn more about.
3. **Find moments of solitude.** I encourage you to reach out to others and build relationships, but sometimes you just need alone time. During that time, give yourself space to think creatively so you'll have ideas to discuss with your team. Research has found that creative people need alone time to come up with new ideas that they can later collaborate on with others.[22]
4. **Have walking meetings.** During every season except winter, the majority of my meetings take place while I'm out walking instead of in a conference room. By walking and talking, I'm changing my environment, which brings out some of my best thinking. Researchers at Stanford University found that walking meetings increase creativity levels by up to 60 percent.[23] Whether you're walking indoors, outdoors, or even on a treadmill, taking yourself out of the normal setting can open up your mind to new possibilities.
5. **Block off time to think.** You can call it "thinking time," "creativity time," or something else, but the point is that sometimes you need to arrange your calendar so that you have a set period every day or week to do nothing but think. You can't rely on other people to give you time to be creative; you need to be accountable.

6. **Travel to new destinations.** If you work for a global company, try to work from a different country at least once a year. If you don't, take a business or personal trip. Through my travels, from Japan to Brazil, I've encountered people, new experiences, and art that have not only shaped my views but also given me new ideas that I take back to my work.
7. **Seek challenging tasks.** When there's no set solution to a particular problem, invent one. Taking on a challenging task gives you no choice but to be creative. The pressure of the challenge could spark ideas for a new process or could open the door to hiring a new employee to help you solve a problem that appears unsolvable.

Seven Ways to Optimize Your Productivity

To maximize your productivity you need a holistic approach, which involves taking a step back and thinking carefully about how you're currently allocating your time, the physical environment that you're in, and your daily team interactions. By doing this, you'll uncover areas that are preventing you from being productive, such as not delegating enough work or having endless email exchanges with team members. The following are several ways to optimize your time, space, and connectivity so you can be the most efficient leader possible.

1. **Optimize your work environment.** Pick a place where there are few or no distractions. If you're working in an open office and it's too noisy, move to the cafeteria (assuming it will be quieter there) or book a conference room for at least the start of your day. If you have your own office space, shut the door during your most productive hours so that you can focus. When you're working, disable your devices' alerts and make sure that your desk area isn't messy. Since your team's overall productivity is affected by each individual's productivity, be

sure to get feedback from all your employees about their work environments.

2. **Optimize your workload.** People are unproductive when they don't know how to prioritize work, a skill that, according to my research, is one of the keys to being a successful leader. If you can't prioritize and delegate tasks, you'll be overloaded with work that you shouldn't be doing. Be clear about exactly what you need to do to achieve your daily or weekly goals, and write it all down. Then rewrite your list, this time putting all the items in order from most important to least. Next to each item write the steps required to achieve it and when it needs to be completed. If there's an item on your list that you know will be a real time and energy suck, delegate it to someone on your team who you know could do the job. As a leader, you should be focusing on the high-impact parts of the tasks, like presenting to other leaders and coming up with new strategies and tactics. "Ambitious team members enable my productivity by giving me confidence in their abilities and allowing me to delegate," says Chris Gumiela, vice president of marketing and advertising at MGM National Harbor. "Simply by taking on more of my workload, teammates give me more time for expansion of my responsibility and to put more time to the tasks that deserve it."
3. **Optimize your flow time.** Flow is the feeling of being completely absorbed in what you're doing and enjoying yourself so much that nothing can distract you—well, almost nothing. What gets in the way of flow are the countless meetings and messages that are competing for your attention and constantly interrupting your day. If you think about it, I'm sure you can recall some occasions when you were in flow. And if you think about it more, you'll probably notice that there's a pattern—for

example, that you're able to slip into flow only right after lunch; it just doesn't happen any earlier or later. Whenever your flow is, block off that time period in your calendar and tell everyone on your team not to schedule meetings then. Flow is easier to achieve when you have a set of clear goals, have the right skills, and work in an environment that's well suited for you.

4. **Optimize your team.** Once you have your own routines and habits set up and you know where you want to go and what you want your team to accomplish, do everything you can to ensure that each person on your team is operating at full capacity. When your team isn't productive, you aren't going to be as productive as you could be. A lot of optimizing your team is about identifying their strengths, weaknesses, and current workloads. If one of your team members is overloaded and stressed out, take away one of their tasks and assign it to someone else. When team members are overwhelmed by one task, they won't be able to do a decent job on any of the other tasks they're supposed to be doing, because they'll be rushing through them so that they can get back to whatever is overwhelming them. If you can't reassign tasks to others or you don't have a team, offload some of the more tedious and routine tasks to a freelancer or a temp worker. "Often our output is a presentation in PowerPoint or Keynote," says Sarah Unger, vice president of marketing strategy, trends, and insights at Viacom. "However, I'm not a PowerPoint designer, and presentations can take forever to create. Hiring designers gave me time to focus on the priority things, while letting designers handle the part where they add the most value."

A number of leaders I interviewed told me about things their own bosses had done that made them more productive. Paolo Mottola, senior manager of content marketing and co-op managing editor at REI, says, "A manager makes me more productive when they give me the room to be entrepreneurial and create work. I'm at my best when I'm empowered and accountable." Sharmi Gandhi, senior vice president of business development at Mic, had a boss who taught her an important trick to being more productive: "Slow down! There is a tendency to think that moving through tasks quickly is the definition of being productive," she says. "However, that can lead to mistakes down the line, as you may not give decisions enough consideration before making them or may not set up a new initiative properly to give it the best chance for success."

5. **Optimize work breaks to maximize your energy.** Fifty-seven percent of office workers take thirty minutes or fewer for lunch, and almost a third take fifteen minutes or fewer. Besides a break for lunch (assuming you actually take one), how many other breaks do you take? Snack breaks, regular bathroom breaks, walks outside, coffee breaks, and all the other kinds of breaks you can think of can be helpful in splitting up your day, giving you some rest, and helping you refocus when you get back to your work. When I was writing this book, I could never work longer than three hours at a stretch without a break. I recommend that you give yourself at least six breaks each day and that you try to plan them out and make a commitment to taking them. If you're one of those people who can work fifteen hours straight without taking so much as a sip of water, force yourself to take breaks. You may feel you're getting a lot done (and you might be right), but when (not if) you burn out, none of it will be worth much.

Sam Violette, manager of e-commerce, mobile, and emerging technologies at Land O'Lakes, Inc., takes his biggest break at the end of the day, and he often spends it shooting hoops. "Some may call breaks meditative or relaxation time," he says, "but I think of it as time to let my brain stop churning on my task list. I'm always sharper and more able to focus coming out of these periods." Adam Miller, product marketing manager at DELL EMC, uses a more structured approach called Pomodoro that recommends determining the task, working on that task for a timed twenty-five minutes, and then taking a five-minute break before repeating—ideally working on a dissimilar task. After four Pomodoros, or about two hours, you take a longer break. "The rationale behind completing dissimilar tasks in repeating Pomodoros is that I can think about a different topic and then return to the previous topic with a fresh perspective," he says.

6. **Optimize your time so you don't waste too much on technology.** If you want to get an accurate picture of how you're spending your time online, rescuetime.com will tell you, to the second. Once you've recovered from the shock of seeing how much time you're wasting, see what you can do to redirect some of that time either toward getting your job done or taking a break. Wasting time online can cost you time and money and will distract you from building the relationships with your teammates that will improve your productivity.

As discussed throughout this book, those relationships are critical to your team's success. "Having strong relationships with my team has driven better outcomes and made processes smoother," says Rosie Perez, lead financial officer for global consumer services business planning at American Express. "Sharing ideas in a room full of strangers or distant colleagues is much more challenging than sharing ideas with a group of people you trust and know personally. In my experience, teams that feel comfortable together tend to take more risks when it comes to sharing ideas, output, and information. In addition, strong relationships encourage a personal commitment to one's colleagues and the collective success of one's team. This is critical in driving creative solutions to problems and getting things done."

The leaders I spoke with had a variety of strategies to optimize their calendars for maximum efficiency. Honeywell's senior director Kiah Erlich schedules hour-long "protected times" that are devoted to doing work on her own. "If you don't protect your calendar, people will consume your sanity and productivity with meeting after meeting all day, every day," she says. Sam Worobec, the director of training at Chipotle Mexican Grill, maximizes every minute of his workweek by auto-scheduling time for others—and himself—weeks in advance. "One-on-ones with my direct reports are auto-scheduled every four weeks, one-on-ones with my employees' direct reports are auto-scheduled every 12 weeks, lunches or coffee with other department heads are auto-scheduled every four weeks, people development meetings with my direct reports auto-scheduled every four weeks, and so on," he says. If you need help managing your calendar, don't be shy about asking for it. "I benefit from having an unbelievable relationship with my executive assistant, Ashley Goodwin, who has made my life so much easier," says Kyle York, general manager of the Oracle Dyn Business Unit at Oracle. "She drives my agenda, which frees me up to worry about the task at hand, knowing she's always going to put me in the right place at the right time. This relationship has been formed over many years of working together, but it is something I highly recommend any executive to invest in. It is worth its weight in gold."

7. **Optimize your meetings so you waste less time and accomplish more.** When Workfront asked employees what most gets in the way of work, 59 percent said wasteful meetings; excessive emails were next, at 43 percent.[24] Most meetings feel wasteful because they're either too long or there's no set agenda. If you want to lead a successful team meeting, create a set of objectives and email it to your colleagues before the meeting starts. Also, when you send calendar notices to block out time for meetings, make sure they're for only half an hour. The tight time frame will put pressure on everyone at the meeting to stick to the agenda and will minimize distractions. If possible, hold your meetings in a different room or venue each week. Changing the environment can stimulate creative ideas and will go a long way toward keeping your teammates' thinking from becoming stale and repetitive.

These seven ideas have been field-tested by experts, and most people have found them quite helpful. However, you know yourself better than I do, so if you have another approach, go for it. Jill Zakrzewski, customer experience manager at Verizon, has a completely counterintuitive approach, but it works for her. "Every day, I sit in a common area with a lot of foot traffic. This means I'm constantly interrupted by people who are taking a coffee break and want to chit chat," she says. "Sure, I often end up staying late on Fridays to finish all the work that was disrupted by these friendly conversations. However, I've built such a network from these interactions that I'm hyper-efficient at identifying the right people for a project, and they respond positively to helping me because of our existing relationship."

Key Takeaways from Optimize Your Productivity

1. **Reduce your digital distractions.** Stop checking your phone every other minute to see whether there's an alert. Turn off those alerts and reduce the number of apps you're running on your phone so you can better focus on your work, get more done, and reduce your stress.
2. **Measure and help improve your teammates' productivity.** Their success reflects on you. If you notice that they're wasting time or are suffering from burnout, redistribute some of their projects to others. The more efficiently they can work and the better they can optimize their own time, the more productive you'll be.
3. **Stop trying to multitask, and quit being a perfectionist.** Instead, perform one task at a time and do it to the best of your ability. Then move on to the next task. No person or project will ever be perfect, so once you're happy with a given result, move on. The feedback you receive from your team and superiors will help you improve the project anyway.

Chapter 3

Practice Shared Learning

School is just the place where you learn the rules of the system. Your life is where you get your education.

—Trevor Noah[1]

The greatest challenge for professionals in today's society is to stay relevant in a world that's constantly changing. The amount of new information that's created every day is astounding, and it's practically impossible to keep up. But what's even more astonishing is that the "half-life" of skills is now only five years. In other words, the skills you have today—and that your employer values—may be close to worthless by the time you get to your next job.

We should all strive to be shared learners. I've taken online courses, but the experiences I've had in classrooms have been a much bigger factor in my education. I have participated in study groups and have been mentored, both of which are in-person experiences that have helped me learn much more than simply reading an article or two. If you care about your team's success, you need to become a shared learner who is open to giving your teammates the knowledge they need when you receive it. At the same time, you need to be just as open to learning *from* your teammates. This free flow of information is good for everyone—and for business. An organization is slowed down if information isn't being shared openly, and when an employee leaves, their knowledge goes with them unless it has been already shared.

By collaborating with your team, sharing what you know with them, and learning *from* them, you'll all acquire information more quickly, retain it better, and be able to apply it in new ways. In one study, Anuradha A. Gokhale, an associate professor at Western Illinois University, found that students who participated in collaborative learning performed better on critical thinking tests than those who studied in isolation.[2] In another study, workers who were in close physical proximity to one another performed about 15 percent better than those who were separated from their colleagues. In fact, the farther apart they were, the more isolated and unhappy they felt.[3] The bottom line is that when you're physically close to other people, you're better able to learn from one another, and you'll all be more productive. This is especially true if you sit next to a diverse population of people who have a strong work ethic and subject matter expertise.

An important note: I recognize that in today's global economy, in which business is operating 24/7, a company's workforce may be spread out in different buildings on the same campus or in different cities or different countries. And I realize that getting all those people together in the same facility (if that were even possible) would be prohibitively expensive. I'm not suggesting that employees of these large or global companies are doomed to be lonely, unhappy, subpar performers. One way to address their physical separation is to use videoconferencing to keep your teams connected. Being able to see one another—even if you're thousands of miles apart—makes for much better communication and collaboration than text messages or Facebook groups. (But don't underestimate the importance of social media. In the course of writing this book I interviewed hundreds of people who were spread around the country and the world. To facilitate the process, I set up a Facebook group. You can't imagine how thrilled I was when two people in the group—John Mwangi, vice president of information governance, law, and franchise integrity at Mastercard, and Jennifer Lopez, senior director of product management at Capital One—discovered that they work in the same building and decided to meet for lunch sometime. You can't get more human than that!)

A shared learning culture is driven by an open network in which teammates gain access to the rest of the team's thoughts, analyses, and resources wherever they are. People learn on their own time, use their preferred devices, and have their own training preferences. Teams need to take advantage of the "always on" team members by providing them with resources they can always tap into. Whether in the form of brainstorming or draft documents, members must make their intellectual capital available for the rest of the team so that everyone is on the same page as often as possible. There are many sources that members can tap into that can help the team foster a strong "knowledge network," including massive open online courses (MOOCs), outside speakers, company-provided education, magazines, online training courses, and college courses.

Be a Shared Learner

Being a shared learner is about identifying opportunities to provide information, resources, and training to team members who need them before they even ask. By paying attention to the skills and information your teammates need (or request) help with, you can deliver the right content at the right time to solve their problems. When you proactively help them develop, they'll want to return the favor and support your own learning. As you do this more within your team, you naturally create a shared learning culture, in which everyone is constantly learning and sharing. The following chart lists a few examples of how to be a shared learner.

Situation	What to Share
You know your colleague is interested in trends in the use of data in HR.	You spot a new white paper focused on people analytics and share it with your colleague.
Your colleague expresses frustration about a new database system she has to learn to do her job better.	You either train her, if you have experience using the system, or you send her a link to a course or tutorial that can help her learn how to use it.

Situation	What to Share
Your team doesn't seem that up-to-date on what's going on in the industry you're in.	Have your team members subscribe to industry-specific online news websites; get a membership to an industry association and have them attend regular meetings.

Shared Learning Exercise

Sit down with one of your employees and pose a problem that you'll try to solve together. The problem needs to be something that affects both of you and is big enough that you really care about it. Have your employee write down all the knowledge, skills, and resources they have (or can find) to help solve that problem. You do the same. Once you're both done, share with each other. Then create an action plan together that will allocate your collective resources to solve the problem, bringing in the rest of your team as needed. This exercise should give you and your employee a clear understanding of the value of shared learning.

Everyone on the team must commit to sharing articles, courses, and resources. When you're hiring new teammates, being a shared learner is a skill and habit that you should look for in candidates.

Overcoming Barriers to Sharing

Being a shared learner requires you to be more aware of your learning style and your willingness to share with your team members. Let's take a look at the obstacles that often get in the way of building stronger relationships and how to address them.

Remove Your Ego

Ego causes us to make poor decisions as leaders. Instead of sharing, we hold onto information, believing that it will allow us to advance beyond our peers. (The truth is that when the team succeeds, so do the individuals who make up that team.) Ego makes us have fewer conversations with others and share our ideas less often because we're afraid of being wrong, of sounding stupid, of being ridiculed, or of giving away information that someone else might be able to use to advance their own career. You need to get rid of your ego and be more willing to take risks or even fail. Instead of thinking of your own career, think about how you can become the provider of information for your team. Your employees will be more productive, which will help you accomplish more. "Knowledge is not job security, nor is it power," says Heather Samp, managing director of network and fleet strategy for American Airlines. "Sharing your knowledge allows you to move to other challenges and spread your expertise in different ways."

Be Less Complacent

We often resist sharing because other leaders and teams aren't actively doing so. You might have a corporate culture in which people work in silos and each individual is learning on their own. It's easy to accept the status quo. Even when we want to change, we tend to retreat to our old ways of doing things (like sitting at a computer taking countless online classes and not telling people about what we're learning).

Be Aware of Different Learning Styles

Everyone has their own learning style and needs. When leaders try to serve everyone in the same way, some people are left out, as not everyone can comprehend or digest the information that way. When you fail to consider the preferences of your teams, you can't deliver for them. To do a better job recognizing differences, you need to show empathy. Get to know your employees more by asking them about their learning style and needs so you can better serve them. Even if everyone on your team needs to learn the same specific skill, they might do so differently, and things will work out a lot better if you take that into account.

Use Technology Wisely

In large companies with distributed workforces, there's a tendency for silos to develop. As discussed in later chapters, these silos interfere with teamwork, communication, and, as you might imagine, the free flow of information and knowledge. Technology—in the form of videoconferences and other online learning opportunities—can create an environment in which employees around the world can actively contribute to the entire knowledge base of the company.

At the same time, technology can be a barrier to good leadership because it's easy to think it's doing our jobs for us and to forget that we're the ones responsible for making it work. Human connections push people to be more open to collaboration.

By having a phone call or a meeting every week, you can make sure that everyone has access to all the resources and training materials so you can have open discussions about what's working and what's not. When we have team meetings, we have a set agenda that includes a list of our priorities, the time allocated to each one, and a series of breaks. This allows us to come prepared for the meeting and focus on the activities that will have the biggest impact on our company. While we use technology to set our calendars and book conference rooms, we (generally) turn off our phones during the meeting so we can be physically present and attentive. Technology can get everyone in sync, but never let it be a barrier to the critical discussions that will most impact your company and career.

Create a Culture of Shared Learning

As a leader, a big part of your job is to get people to help one another learn and establish a culture in which everyone is accountable for the team's success. If each team member can actively train the others in new skills, the team will stay relevant together and will be more productive and successful as a result. The most successful leaders of the present and future are those who embrace and actively practice shared learning over self-learning. When you're actively helping your teammates, you become a role model for how they can better learn and support one another. The following is a list of recommendations for creating a culture of shared learning.

1. **Ask for and give feedback.** By giving your employees regular feedback and then soliciting theirs, you'll create an environment in which it's acceptable to both criticize and compliment, and you'll facilitate invaluable conversations that will benefit everyone.

Ask These Questions in Your Feedback Session

- What's preventing you from sharing resources?
- How can I best support your ongoing development?
- What sources do you turn to first when trying to learn a new skill?
- What publications or news sites do you read every day?
- What do you think is stopping the flow of information in this team?

---

2. **Track accomplishments.** Take a hard look at what you and your team have done over the past several months and examine the actual business results of your activities. Think about the team's accomplishments and employees' individual ones. If some team members aren't contributing as much or lack certain important skills, you need to get them up to speed. By identifying gaps and weak points, you can create a learning ecosystem that will support everyone.
3. **Be flexible.** As mentioned at the beginning of the chapter, change is constant, and as a leader, it's on you to ensure that your team adapts to that change. As you or your teammates are researching new trends, skills, and potential market opportunities, instead of keeping information

to yourselves, share it immediately. But be flexible with how you share new material. Having a proper mix of face-to-face meetings, videoconferencing, email, and social media can support everyone's needs.

4. **Have a positive attitude.** Push your ego aside and get excited about improving the lives of those around you. When you're in shared learning discussions, encourage—and embrace—criticism, because that's how you'll get the most honest feedback possible. When you hire for your team, look for those who have a positive attitude about helping others, and be wary of those who seem to be focused on becoming the next CEO.
5. **Promote the expertise of others.** Everyone has their own unique skills and can be a teacher, not just a learner. Over time the interactions you have with your teammates and the actual work they produce will let you know what their strengths and weaknesses are. Pay special attention to what they're good at, and when you see an opportunity for them to help, bring them in.

The Importance of a Shared Learning Culture

When you and your team are sharing knowledge freely and openly, you are building a sustainable culture that the entire organization can benefit from. When we asked managers and human resources (HR) executives how they preserve and strengthen their workplace culture, over two-thirds said through training and development programs.[4] These can help you increase productivity and efficiency because your employees will have the knowledge they need to accomplish your goals and theirs. Sharing information and skills can also help you increase employee satisfaction. We have found several times that when recruiting and retaining employees, pay and health care are only conversation starters; what employees desire more than anything else is training that can help them advance in their jobs and their careers. While some of your employees may have different ambitions than others, they all care about their careers. They know (and hopefully so do you) that if you aren't becoming better at your craft, you won't last long.

Ask anyone in HR, and they'll tell you that it costs a lot of money to replace an employee. A shared learning culture in which employees are always acquiring new skills and honing existing ones increases loyalty and decreases turnover. It also improves morale by developing employees into mutual supporters and champions for one another. It's important to move them away from a typical "winner-takes-all" or "every-person-for-themself" mentality and toward one that embraces collaboration and helpfulness. Workers who become good at shifting back and forth between teacher and student are humbler and have fewer ego issues.

How to Teach Your Employees Something New

Part of creating a shared learning culture is understanding how to teach skills. On-the-job training is the best way to learn, and in-person teaching can be incredibly powerful in helping you build a stronger relationship with your colleagues. Here are some ways to teach others a new skill.

1. **Empathize with your coworkers.** Because you know something that they don't, you're the authority figure in the teacher-student relationship. To make them feel more relaxed and comfortable working with you, consider sharing a weakness of your own or a skill that you could improve on.

2. **Display your skill.** When you're showcasing your skill, explain the step-by-step process you use so your colleagues can follow along with you. For instance, if you're showing them how to use a computer program to create a short piece of code, walk them through the process of how you got to the final product so that they can replicate it on their own.

3. **Encourage them to practice the skill.** This is especially important for hands-on learners who need to perform an action several times to master it. After you walk your teammates through how you apply a skill, let them test it out on their own to see whether they can repeat the process you used and achieve the same or a similar result.
4. **Give them feedback.** Once your coworkers have attempted to complete a task using the skill you taught them, review it. Explain what they did right and how to improve. If they're having trouble, go back to step 2 and review your process again. Some people take longer than others to learn and master a new skill, so be patient.
5. **Follow up.** After a week or two, have another meeting to see whether your teammates have been able to successfully implement the skill you taught and to answer questions or provide additional help. By checking back regularly, you're ensuring that your teammates will improve, and you're demonstrating your commitment to them and their development and success.

Another reason to teach others is that, in my experience, it's the best way for you to learn. Many of the people I interviewed for this book agreed. "Training another employee not only helps them to do their job but often helps the trainer be better at the job," says Leor Radbil, senior associate in investor relations at Bain Capital. "When you do something all the time, sometimes you gloss over steps or just get used to the routine. When training someone, you go through the process methodically. The trainee may often teach the trainer a thing or too as well. The trainee may ask questions that hadn't occurred to the trainer, which lead the trainer to learn new things as well. And of course, sitting with someone and allowing them to see the process for the first time with fresh eyes can add a new perspective. A creative trainee won't just absorb the information and regurgitate it, but will look for ways to improve the process or make it more efficient."

Heather Samp of American Airlines told me a story about a young man who had what she considered to be outstanding leadership potential. She thought he'd make a good manager for one of her teams and set about teaching him everything she knew about that area of the business. A few months later she beamed with pride as he answered questions and thoughtfully questioned others in an important meeting. "At that moment, I knew we both had accomplished something. He accomplished the goal of learning the business, and I accomplished my goal of making myself replaceable," she told me. "This moment was a transition in my career where I have found less satisfaction through my own accomplishments but rather finding it through how I can help others achieve their goals. It is a truly rewarding feeling unlike anything I have felt before."

How to Learn from Others

A lot of people—especially those in leadership positions—have the idea that asking for help is a sign of weakness. That applies to both education and training. "It's as if you aren't capable of succeeding on your own," says Bill Connolly, director of content at Monotype. "Success isn't a zero-sum game. When employees are willing to both help others and ask for help from others, everyone can achieve at a much higher rate. I seek out training and personal growth opportunities whenever I can and am never afraid to ask for feedback or support on a project if I believe it can benefit from a different perspective."

If you're young and still at an early point in your career, there's no shame in learning from your teammates. "I work with a lot of older people who have worked at Cisco for 10 to 20 years and are extremely knowledgeable about supply chain," explains Caroline Guenther, integrated business planning manager at the company. "I learn from them every day, and they are excellent teachers, which is part of what makes Cisco such a great collaborative environment."

If you're more established in your career, recognize that you're surrounded by industry experts, many of whom may be younger and possibly further down the organizational chart than you are.

"Your colleagues will teach you far more than any college," says Kiah Erlich, a senior director at Honeywell. "I am learning from my team how software is coded, how jet engines keep an aircraft in the air, and how satellites bounce waves of Wi-Fi connectivity to keep you productive even while flying. Thanks to the experts I work with, I am getting smarter every day and more capable of making better business decisions."

How to Sustain a Shared Learning Culture

As a leader, you need to create the right values, processes, and practices that will encourage people to share information and learn from one another. The goal is to increase everyone's shared intelligence and skill level, making all employees more productive and satisfied with their jobs. Doing this might pose a challenge, because some of your colleagues will be more inclined to share only when they're forced to, at least at first. Here are suggestions for creating and sustaining a culture that embraces shared learning.

- **Hire collaborative employees.** When you evaluate candidates, make sure that you spend part of the interview trying to assess how they feel about the importance of learning. For instance, you might ask something like "How willing are you to teach a new skill to a fellow employee?" or "Tell me about a time when you helped an employee complete a task or project that wasn't related to your job." The answers to these questions will give you a better sense of how willing an individual is to share knowledge. You want to hire people who are intellectually driven and have an enthusiasm for the learning process.
- **Create a formal training plan.** The best way to get all your employees on board is to have a mandatory program that everyone supports. Instead of creating the program by yourself, incorporate everyone's ideas. This will make them feel more involved and will increase the likelihood that they'll execute on it. You want your team

members to take the program seriously, so explain how it benefits them as individuals, not just the team. I recommend going into detail about how their colleagues' performance can reflect their own and how people will support them if they help others first.

- **Recognize it when you see it.** When you see an employee helping another one, say something positive. If one employee is teaching another a new skill and it's benefiting them both, applaud the effort. In addition, you should reward those who are investing in new skills and abilities outside of office hours; others will copy that behavior.
- **Build custom learning paths.** Have individual conversations with each of your employees to figure out what skill areas are currently required for their jobs, and talk about what they'll need to learn to be successful in the future, both near and long-term. Helping your subordinates understand the requirements of different positions within the company is an incredibly important part of being an effective leader. By setting realistic and reasonable expectations and being straightforward about the skills that are needed, you'll be setting up your employees for success. They, in turn, will be more loyal to your company and more committed, making it better.
- **Learn, learn, learn.** It's essential that you and everyone else on your team keep up-to-date on new developments in your industry and, more important, that you share that information. There are many great ways to do this; here are just a few. Rashida Hodge, vice president of Watson embed and strategic partnerships, reads industry journals, magazines, books, and blogs. Paolo Mottola, senior manager of content marketing and co-op managing editor at REI, listens to business and industry podcasts. John Huntsman, associate director of information and data management at Bristol-Myers Squibb, is a big consumer

of professional market research (Gartner and Forrester) and industry news digests (Fierce and Pink Sheet). Jennifer Schopfer, vice president and general manager of transport logistics at GE Transportation, brings in external technology and industry experts to educate her team on external trends. Team members also attend conferences and trade shows. Chris Gumiela, vice president of marketing and advertising at MGM National Harbor, does many of these things but ultimately prefers to engage in conversation with likeminded individuals on relevant topics. "This is where perspectives come out and debates can be had in a healthy fashion without any potential for repercussion," he says. Ilona Jurkiewicz, vice president of talent and development, early careers, at Thomson Reuters, sets aside ten minutes at the end of each day (sometimes at her desk, but sometimes on her commute) to consider what she's learned that day and whom it could benefit. She then sends articles, quotes, connections, and new ideas to her network. "This not only reinforces my own learning but allows me a systematic way of sharing resources, and it helps me maintain my network." And finally, Tracy Shepard-Rashkin, sustainable communities brand manager at Unilever, started a quarterly lunch and learn last year in which she takes the coolest case studies she's learned at conferences, does additional research, and over a meal shares this information with more than one hundred marketers she works with. "This quickly became not only one of my favorite parts of my job, as it allowed me to share things I was very enthusiastic about with a broader group, but it ended up being a great benefit to me personally: My colleagues started thinking of me as the go-to person to share interesting articles or presentations with, in the hopes they might be the focus of the next quarterly lunch and learn!"

Shared Learning Closes the Generational Divide

There's a great cultural and technological divide between younger and older workers, but both can benefit from each other's knowledge and skills in important ways. Whereas older workers have had years of experience, younger ones are more likely to have different perspectives from growing up in a very different time period. Older generations have benefited from in-person education and on-the-job training and know the value of in-person meetings. At the same time, they may not be as technologically adept as younger people. And while those same young people may have collected a lot of trophies and ribbons growing up (admittedly, some of which were just for showing up), they also have learned about the power of social media and how to use it to connect with people of diverse backgrounds all over the world.

What Younger Workers Can Teach Older Workers	What Older Workers Can Teach Younger Ones
• New technologies that will impact internal collaboration and their profession and industry and how to use them. • The importance of diversity and how it can benefit the team, since younger employees are the most diverse in history. • How change is inevitable, why the skills of today may not be as valuable in the future, and how to learn new skills. • Why they shouldn't give up on their dreams. Research shows that younger workers are more optimistic and can use that to inspire older workers. • The collaborative mind-set that will help older workers best interact with them, brainstorm, and come up with new ideas.	• The struggles and setbacks of building a career and the importance of having years of experience. • The soft skills that have helped them build the relationships that have made them successful. • The loyalty that makes others on your team want to invest in your learning and development. • The regrets they might have had in their career and how to not make the same mistakes. • How to manage corporate politics that naturally occur in any corporation, especially larger ones.

What Younger Workers Can Teach Older Workers	What Older Workers Can Teach Younger Ones
	• The skill to handle conflicts in the workplace and the wisdom to use those conflicts to actually solve problems and form stronger relationships in the aftermath.

Crossing the generational divide can advance your career and make it easier to manage older teammates. Think of this as a mutually beneficial learning situation that will help bridge those relationships in a positive way. Jessica Latimer, director of social media at Alex and Ani, admits that she has colleagues who still don't have social media accounts, or if they do, don't understand how to use them. "I actually get excited by this and choose to see it as an opportunity to educate them and potentially propel them into joining a network," she said. Thanks to her efforts, her colleagues benefit from staying relevant, while she has more advocates for her program—a win-win situation!

Although there are plenty of generation-related differences, as a team we have the common goal of performing our jobs, generating business results, and hopefully building strong bonds along the way. That's why workers of all age groups need to come together and focus on the mission: fostering a culture in which everyone can continually learn and improve.

Key Takeaways from Practice Shared Learning

1. **Help your team out without asking for anything in return.** The more you invest in helping your team learn and develop, the more successful everyone will be. Instead of hoarding information, share it so team members have all the resources and skills they need to meet your needs and achieve results on their own individual projects.

2. **Take the time to understand your employees' learning styles.** This will help you tailor your approach to their needs. Sit down with the employees separately and get to know what resources they turn to when learning a new skill and how you specifically can lend your support to their development.
3. **Focus on building a shared learning culture.** Help ensure that all employees on your team or in your entire company are openly sharing with and helping one another. This culture will help you and them achieve all your goals because, as the old saying goes, "a rising tide lifts all boats." With the half-life of a skill getting shorter and more industry disruption, it's imperative that you become a shared learner to keep up with your business's demands.

Create Team Connection

Chapter 4

Promote Diverse Ideas

It is not the manager's job to prevent risks. It is the manager's job to make it safe to take them.

—Ed Catmull, CEO, Pixar[1]

Not long ago, when people talked about diversity they were referring to a number of visible demographic attributes, such as race, ethnicity, age, and sex. After a while the definition of diversity expanded to include less visible attributes such as sexuality, religion, and even educational attainment. Today diversity has become even broader and now includes intangible characteristics such as upbringing, socioeconomic status, life experiences, and worldviews. Take a look at the following chart, which is a by-no-means-comprehensive list of the many factors that now constitute diversity. Feel free to add your own.

Types of Diversity	
Race/Ethnicity	white, Hispanic, African American, Native American, Asian
Education	no schooling, high school diploma, associate's, bachelor's, master's, professional, doctorate
Gender	male, female, nonbinary, transgender, gender fluid

(continued)

Generation/Age	silent generation, baby boomer, Gen X, millennial, Gen Z
Employment Status	freelance, full-time, part-time, in-office, remote
Religion	Christian, Muslim, Jewish, Catholic, Buddhist, atheist, agnostic
Politics	Republican, Democrat, Independent, Libertarian, Green Party
Sexual Orientation	heterosexual, bisexual, homosexual, polysexual, asexual, demi-sexual
Profession	marketing, engineering, operations, finance, accounting, and so on

One might reasonably think that with our ever-expanding definition of diversity and society's ever-changing demographics (which to a great extent reflect that diversity), our companies would be supporters of inclusiveness. Unfortunately, although many companies—including Facebook, Apple, and others in Silicon Valley, which is home to some of the most innovative companies in the world—say that they value diversity, when we look at their workforce composition, it appears pretty much as homogeneous as it always did. Let me give you a few examples.

- In Silicon Valley, for example, African Americans and Hispanics make up only about 5 percent of the workforce.[2] These employees are often stereotyped, discriminated against, and passed over for promotions, which all too often leads them to quit. It's no wonder half the employees at these companies believe that major improvements need to be made in this area.[3]
- There is an ongoing global conversation about women's rights and sexual harassment in the workplace, which has given rise to employee resource groups, events, and

conferences. Sheryl Sandberg created the Lean In movement to encourage women to seek challenges and pursue their careers, creating circles to gather in and support one another. Nevertheless, women fill a mere 24 percent of senior business roles globally[4] and a paltry 4.2 percent of CEO positions in Fortune 500 companies. Just a few years ago, I spoke at a women's conference with an attendance of over ten thousand. For once I felt like the minority!

- Over the years countless articles have mocked the millennial generation. We've been called lazy, entitled, narcissistic, and unfocused—which is pretty much the way younger generations have always been stereotyped by older ones. The reason we can't shake these horrible, and mostly untrue, stereotypes is that the media (and social media) amplifies them. A study conducted by Wharton marketing professor Jonah Berger found that the most popular articles in the *New York Times* are those that elicit anger from the reader.[5] The more articles published that bash my generation, the more traffic media companies receive and the more advertising money they bring in. Don't believe the stereotypes! As of 2015 millennials were the largest generation, not to mention the most racially diverse one in history, with 43 percent being nonwhite.[6]
- The pressure to get a college degree has always been strong and seems to be accelerating. In 2016 nearly a third of all Americans had at least an associate's degree, 15 percent had a bachelor's degree, and 6 percent had a master's.[7] By comparison, nearly 8 percent of people age twenty-five and over have a master's degree, about the same proportion that had a bachelor's or higher in 1960.[8] (This may explain why some people have started to say that a bachelor's degree is the new high school diploma.) But while most online recruitment tools allow employers to filter out applicants who don't have at least

one degree, a small number of employers have recognized that always hiring people from the same schools who've taken the same courses and have the same mind-set results in a homogeneous, less innovative workforce. These more open-minded companies, including EY, PwC, Ogilvy Group, and Apple, have lowered GPA requirements or are interviewing candidates who haven't attended college at all.

Diverse Ideas Are Critical to Your Team's Success

As I have noted, diverse workforces tend to be more productive and creative. And encouraging diversity can increase employee engagement and boost the financial health of your organization as a whole. However, it would be impossible to create a workforce made up of a perfect, mathematical representation of every conceivable demographic. That said, there's one type of diversity that I think *is* achievable. In a global study, my company asked more than four thousand young workers about the type of workplace diversity they value the most, and they didn't say gender, age, religion, or ethnicity. Instead, they said, "diverse points of view." I call that *diverse ideas,* and I'm a huge fan. Focusing on people's experiences, mind-sets, and viewpoints effectively incorporates diversity in all its forms.

Achieving diversity of gender, age, and ethnicity is relatively easy, says Charlie Cole, chief digital officer and vice president of Tumi. "I think a team of 10 Harvard MBAs would be less effective than two high school dropout coders, two MBAs from Jakarta, two undergrad varsity athletes from Seattle, two art history majors from Atlanta, and two MIT statisticians. And I frankly don't think it would be very close."

Groupthink Is the Enemy of Diverse Ideas

The enemy of diverse ideas is groupthink, which is what happens when your team reaches a consensus that disregards any opposing viewpoints. In 2015 the US Environmental Protection Agency (EPA) discovered that Volkswagen had installed software in eleven million cars around the world (including half a million in the United States) that deliberately created the impression they were safer for the environment. The software was designed to activate their emissions controls to meet US standards only during testing. The rest of the time, they were emitting up to forty times more than the allowable limits. As a result of this discovery, Volkswagen had to spend more than $18 billion recalling vehicles and fixing the emissions issues[9] (and that doesn't include the billions in fines the company has had to pay). The root cause of Volkswagen's dishonesty is its corporate culture, which is dominated by engineers who decided—apparently in the absence of opposing viewpoints—to engage in this scheme. The company operated like an oligarchy, with former chairman Ferdinand Piëch's brother, Dr. Hans Michel Piëch, on the supervisory board.[10] The danger of groupthink is that without diverse ideas, organizations too often make bad decisions that lead to financial loss and harm—not just to the company but to their customers and, in this case, the world's environment.

"Diverse ideas not only decrease the risk of groupthink but also result in the most innovative solutions and highest productivity over time," says Vivek Raval, head of performance management at Facebook.

You know when groupthink is occurring when there aren't many debates, when you find quick solutions to complex problems, and when anyone who disagrees with the consensus is ridiculed or negatively stereotyped. Other red flags to watch out for are teammates who agree on one decision despite alternative evidence and situations in which no one disagrees or encourages others on the team to try something new (or worse, when people seem afraid to take a position that's contrary to the group's). If your teammates feel they have to follow your guidance because they'll be punished if they don't, you're inhibiting diverse ideas.

Common Groupthink Phrases

"Let's work through this project like we did before, because it worked."

"Our work is exceptional; nothing has ever gone wrong."

"Don't listen to them; they have no idea what they're talking about."

"I know we all agree here."

"We're all aware that this will work out."

Diverse Ideas Lead to Better Business Outcomes

The variety of views that are inherent in diverse ideas can create disagreements and disputes, which—as long as they're handled respectfully—in turn give rise to creativity and innovation, both of which are important traits in high-performing teams and companies. Although some disagreements can be hostile, most are harmless, and many are actually quite worthwhile. Being exposed to different ideas creates a certain amount of tension that engages employees by making them think about their own behavioral styles and contributions and makes them value those aspects in their teammates and others.

Diverse ideas also help you guard against groupthink and prevent overconfident "experts" from getting their way all the time. When everyone feels comfortable bringing something to the table, there's a stronger feeling of connection and safety at work, which is what we all want, right? The positive side effect of identifying and hiring different types of employees is that they'll do the same when they're in management roles; so will the next generation, and so forth.

Your customers come from a variety of backgrounds, so having people on your team who can better understand their diverse customers' languages and views can be extremely valuable. By having diverse teams, you can better meet your customers' needs or even learn what their needs are in the first place. As Facebook's head of performance management Vivek Raval puts it, "Our customers are not uniform in thoughts and preferences, and thus we cannot afford to be uniform in our ideation and execution."

Furthermore, when you have a variety of employees who collectively have dissimilar skills and experiences, you can provide more to the organizational culture. This lends itself to more effective execution, in that employees from various backgrounds can come together to perform at a higher level, which benefits everyone's individual careers and results in higher productivity, profit, and return on investment.

"Consumers buying our client's services are more diverse than ever, and it's easy to get pigeonholed into certain campaign ideas, media targets to pitch, etc. when you don't have people from other backgrounds contributing," says Emily Kaplan, senior account supervisor of DC Brand at Edelman. Emily gave me a great example of diverse ideas in action. "Most of my Starbucks team are Caucasian females, but Jarryd, an African American man, helped us come up with great pitch ideas for men's lifestyle media and formed new relationships with reporters we'd never worked with before. He also introduced us to trends like 'Black Twitter' and explained what this meant for our client and our work."

Ten Signs You Lack Diverse Ideas (Checklist)	
1. You try to control team conversations rather than influence them.	
2. You operate in a silo instead of putting yourself out there.	
3. You view vulnerability as weakness in your teammates.	
4. You hold back ideas because you don't want to be judged.	
5. You leave a teammate out of a meeting because they don't share your same opinions.	

(continued)

6. You focus only on your strengths without accounting for your weaknesses.	
7. You refuse to challenge traditional ways of doing things.	
8. You have an unconscious bias when hiring and working with others.	
9. You are too comfortable in your systems and refuse to change.	
10. You don't research and acknowledge the differences and preferences of others.	

If you find yourself checking off multiple items on this list, you need to start becoming more aware of and thoughtful about embracing diverse ideas. Think about how you can become more open to others and incorporate their thoughts in your decision-making process. Have them review your checklist and do their own. This could create a meaningful conversation between you, with some potential positive outcomes.

Technology Can Inhibit Diverse Ideas

Technology has become the communication platform of choice for most people and in most workplaces. In theory, that's a wonderful thing. After all, regardless of ethnicity, age, or any other factor, technology gives everyone the same access, rights, and privileges. The hope was that technology would help create more inclusive teams. Although that has definitely happened in some places, technology has also created a digital divide between workers, especially those from different age groups. Because older workers didn't grow up with the same tools that younger workers did, it's sometimes more difficult for them to effectively apply those tools. As a result, it may be harder for them to connect with us on a professional, or even personal, level.

Technology was also supposed to put everyone on the same plane and make it easier for people to understand and connect with teammates, regardless of geography, language, or culture. Unfortunately, text messages, status updates, emails, and other types of technology-based communications have caused more problems than they have solved and may actually become a barrier to diverse ideas on your team.

When everyone is relying on all sorts of devices, apps, and messaging services, they're spending less time understanding where others are coming from, and they never have a chance to experience their emotions. When you're sending and receiving messages, you lose the tone, language, and expressions that in person help you get a sense of who people are, not just what they say. While technology platforms may make it easier—and more comfortable—for others to share their ideas and thoughts, people are also hesitant to publish something that might be ignored—or even be used against them.

As we know, although words are important, a great deal of communication is nonverbal; body language and tone of voice play important roles. We've all had plenty of experiences in which we have misinterpreted something we've read or someone else has misinterpreted something we've written—and that's when we're speaking the same language. In fact, a study published in the *Journal of Personality and Social Psychology* found that we think we have correctly interpreted the tone of emails 90 percent of the time, but actually we have only done so half the time.[11]

Imagine how much riskier relying on written communication is when the sender and receiver come from different cultures or one of them is trying to communicate in a non-native language. All too often the result is that teammates misunderstand what's expected of them, do the wrong projects at the wrong time, or inadvertently convey incorrect information. Feelings can get hurt, relationships can be strained, conflicts can arise, and team and company performance can suffer. Technology isn't the solution to bridging different cultures and languages, but you can use it to create more in-person meetings and get a better sense of what people are thinking.

In chapter 2, I referenced a study by Mahdi Roghanizad that found face-to-face requests were thirty-four times more effective than email ones.[12] Sjoerd Gehring, global vice president of talent acquisition and employee experience at Johnson & Johnson, wasn't familiar with Roghanizad's research, but he definitely agrees with its conclusions, particularly in reference to his early career, when he was dealing with direct reports, all of whom were at least ten years older than he. "Getting to know them at a more personal level was key to earning their respect. But not the way I was used to," he says. "My go-to way of communicating was texting, Twitter, and LinkedIn, but that wasn't helping me connect with my new team. Then I took someone to lunch, and everything clicked. We talked about their experience, their background, their family, their passions. Nothing like a little face time (as opposed to FaceTime)." Sjoerd's advice? "It's essential to see things from others' perspectives. Double down on forging personal connections, and learn what motivates your team and how they communicate. You'll become much more effective at leading them."

Imagine this: you and a colleague have been texting back and forth for days, trying to finish a particularly challenging project. But you can't quite get there, and you keep getting pushback from your boss. Probably without realizing it, you've made a conscious choice to not incorporate diverse ideas. Clearly your digital dialogue isn't working, and all you've done is figure out a lot of different ways to not solve the problem. Bringing in a new perspective or two might have been all you needed to succeed. By holding an in-person meeting instead of sending texts, you open yourself up to the opinions of different employees, who have various viewpoints that should be taken into account before you make a business decision that could affect your entire team. Avoid being confined by technology by using in-person meetings or conference calls to incorporate the views of others for a more refined solution.

Other Barriers to Diverse Ideas

Technology can be a major barrier to diverse ideas in the workplace, but there are others:

- **Communication issues.** People may speak different languages or be from different parts of the same country, where the same words might mean something completely different. If you say something to one of your teammates that is misinterpreted, it will take a lot longer for you to get your point across. There are major differences between cultures in how people do everything from handling deadlines and delegating authority to communicating with teammates and resolving conflict. You need to create a safe environment for everyone and take the time to get to know each teammate and their specific needs and styles to effectively communicate with all of them.
- **External resistance.** Your superiors and even your teammates may not support your efforts to implement diverse ideas.
- **Internal resistance to risk taking.** Author Steven Pressfield calls this "the resistance," or the voice in our heads that tells us to be careful. It's much easier to accept the status quo and be complacent than to put your heart, feelings, and reputation on the line.[13] However, by doing the latter you actually gain more confidence and have an opportunity to make positive change. We sometimes suffer from the "we've-always-done-it-this-way" mentality, which makes it impossible to ever do things better.
- **Unconscious bias.** Like it or not, we're all influenced by events, people, the media, and other factors. A Korn Ferry study found that 42 percent of workers believe there's an element of unconscious bias in their workforce when it comes to diversity.[14] Our own biases can turn bridges into walls and harm our work relationships.

How to Effectively Manage Diversity

We can all do a better job of incorporating the ideas of others and making people feel more comfortable at work. To manage diversity effectively, you need to be more open to unconventional candidates, get to know people on an individual basis, create a safe space where people are supported, and reward and recognize diversity when you see it.

1. **Hire unconventional candidates.** To get the right level of diversity in your team, you must change your hiring criteria. Don't just look at candidates' accomplishments and where they went to school. Ask about their passions, who and what most influenced them, and their interests outside of work. From the job description to the interview process, your qualifications should be looser, and you should account for unconscious bias so that you don't end up passing on someone because of what they look like. Just because someone doesn't have a degree, or grew up in a part of the country you've never heard of before, doesn't mean that they couldn't add something special to your team.

Leadership Exercise for Hiring Unconventional Candidates

During the interview process, challenge candidates to brainstorm new ideas for a project you're currently working on as a team. You don't have to mention specifically what the project entails, but try to see how well they think on their feet and whether their ideas are different from what you or your colleagues have come up with. If they're able to constructively criticize your way of thinking or come up with something completely different, they may be able to bring some well-needed diverse ideas to the table if you decide to hire them.

---

2. **Understand individual needs.** Instead of just observing your team from a distance, set up one-on-one, in-person meetings to get to know your teammates better. Don't bother texting or instant messaging them; that won't

give you a sense of their emotions, views, and creativity. Sometimes you need to look inward to get an outward, more general view of things. For instance, one of your employees might be interested in time management, while another couldn't care less. One could be more introverted, while another one could be the life of the party and want to plan every social event. Another may prefer running things by the team before making decisions, while yet another may act first and then ask for feedback afterward. You need to get to know the people you work with and their habits so you can lead them in the most effective way.

Leadership Exercise for Understanding Individual Needs

Sit down with each employee and ask about their strongest values. During the conversation, share *your* strongest values with them. This way, you can both learn more about each other. By learning about each other's values, you get a better sense of what you both care about and how to best integrate the employee in your team, assess their abilities, and provide for them. Take notes during this one-on-one meeting and share them with each afterward.

---

3. **Create a safe space.** When Google managers embarked on a quest to build a more productive team, a project known as Project Aristotle, they interviewed hundreds of employees from throughout the company. They wanted to use data to find the proper mix of employees to identify leaders. What they found was that the best teams respect one another's emotions and are mindful that all members should contribute equally to the conversation. Having a safe environment in which people feel comfortable interacting and sharing their perspectives is what makes a team successful and more productive than

others.[15] Those feelings of safety and security reduce employees' stress and encourage them to share their ideas instead of keeping them to themselves.

Leadership Exercise for Creating a Safe Space

People generally are more willing to share their ideas if they can do so anonymously. When you tackle your next project, have each person on your team write down an idea—but not his name—on an index card. Once you've collected the cards, make a list of the ideas and group similar ones together. Then hold a meeting to discuss the ideas and get feedback from the team. Take note of who publicly takes credit for their ideas and who stays quiet. That will give you a sense of whom you need to spend more time with because they need more comfort and safety.

---

4. **Recognize thoughtfulness.** When you see employees being courageous and open, tell them they're doing a great job. Let them know that idea sharing is encouraged and that the more they share, the better off the team will be. Find a small way to reward this type of behavior—and be sure to tailor the reward to the individual. (Some would prefer public recognition; others would be much happier with a Starbucks gift card.) Incentivizing employees in this way will make them do the same for their employees in the future when they're in leadership roles. But keep these rewards separate from performance-based incentives that reward innovation, process improvement, creativity, or innovations that produce measurable bottom-line results. In a conversation I had with Wharton professor Adam Grant, he advised me to "recognize that dissenting opinions are useful even when they're wrong, and go out of your way to reward them. Promote and celebrate the people who openly disagree with you and criticize you."[16]

Leadership Exercise for Recognizing Thoughtfulness

Use a collaborative or social media platform to encourage employees to share others' ideas or compliment one another. This type of real-time feedback and recognition will give people more confidence, build team camaraderie, and foster a culture of diversity. As the leader, you should be the first to post so that you set the precedent and show your commitment to recognizing others. Hopefully when others see that you're on board, they'll follow your lead.

5. **Communicate relentlessly.** J&J's Gehring told me that one of his first acts in his current job was to hang a big poster outside his office that said, "Ideas over Titles." It sent a message that ideas count more than anything else, regardless of pedigree or seniority. But hanging a poster wasn't enough. "I worked hard to draw those ideas from my team and tap into their vast experience. I reached out relentlessly, early, and often. I created weekly videos to introduce my leadership style and set the tone. I got serious about using social media as a management tool, both internally and externally. Transparent and direct communication will help you earn the respect of your team, especially when combined with a clear vision and sense of purpose."
6. **Encourage people to share their ideas.** Successful leaders around the world do this in a variety of ways. Patricia Rollins, senior director of marketing at CA, dedicates a few minutes of every team call to thinking through what's working and what's not. "I encourage one 'get fired' idea (something that's really out-of-the-box thinking) to help transform our roles. I then assign a lead to that idea to get it executed." Sam Worobec, director of training at Chipotle Mexican Grill, encourages his team

to try out new projects on their own time. "So long as the idea is tied to a problem we are all trying to solve, they are wide open to attempting to solve the issue and bringing it to the team. From this, some of our most transformational ideas have emerged." Ross Feinberg, senior director of strategy and operations at Akamai, has a different approach: "My favorite thing to ask is, 'What do you think?' and I make sure to go around the room to get everyone to chime in. Some people will not be forward with their thoughts, so they need to be directly asked, but often these people have the best ideas." At Liberty Mutual, everyone on Jenna Lebel's team is empowered to take smart risks—even if they end up failing. "We believe that sometimes you win and sometimes you learn, but both are equally valuable to us," says Lebel, who's the vice president of brand and integrated marketing. "And we reward the team accordingly—we reward for successful ideas and for ideas that never make it off the ground. The team takes comfort knowing that they can try new things and creatively tackle business challenges without fear of consequences if things don't go well. We all know that it's okay to fail, but we need to fail fast and take the learnings with us as we go forward."

How to Promote Diverse Ideas in Your Workplace

Aside from the leadership techniques just discussed, think about how to foster an overall culture that promotes diverse ideas. Focusing on your culture can have a long-term positive impact on hiring, managing, and promoting. It can also set up future leaders for success, because the company's DNA contributes not only to the makeup of your employee base but also to the ways people behave toward and think about one another. The following are some steps that will help you create a culture that will support your organization right now and in the future.

1. **Assess your current situation.** Take a long, hard look at your recruitment processes and how (whether) your team and your company support differing perspectives. Conduct an employee satisfaction survey to identify your current diversity challenges and opportunities. Are people supported when they share new ideas? Do they feel that their voices are heard and their ideas are incorporated into new processes and projects? This will give you an idea of the new policies, guidelines, and conversations you'll need to implement to fully embrace diverse ideas.
2. **Seek executive commitment.** Once you discover the obstacles and issues you're facing, write them down and use the data from your survey to make a case to management that things need to change. Once the managers are convinced that there's a problem, get them to commit to supporting your attempt to resolve that problem.
3. **Create an action plan.** After determining the attitudes toward diversity and the issues that your teammates face, it's time to develop a plan. The plan should include your recommendations for solving the problem and a timeline that you can be accountable for. For instance, during the first week you should have a kickoff meeting at which you present the results from the engagement survey and solicit people's opinions.
4. **Measure the results.** After implementing your action plan, issue the same survey to your team to see whether there's been any improvement. Hopefully you'll be able to use the new data to support the case you're making to your executive team: that tackling this problem was worth your time and has benefited the team and organization.

Inclusion Exercise

Take a stack of colored note cards and give one to each of your team members at your next meeting. Think of these as Get Out of Jail Free cards from Monopoly, except that they're actually Take a Risk for Free (TARFF) cards. The goal is to encourage your employees to share a new idea, meet someone new, or do something else that might seem a little (or a lot) intimidating, while making it less scary to do so.

The caveats with these TARFF cards are that (a) they must be used by the end of the quarter, and once used, the card has to be turned in, and (b) if you don't use the card during the quarter, it will impact your salary or bonus. Having a penalty attached to not performing forces your teammates to act, which is the intent.

How to Manage Different Diversity Situations

As a leader, you'll undoubtedly have to deal with a variety of situations in which you need to manage diversity. Before you jump in, it's important to step back and get a clear view of what's really going on, understand where the individuals involved are coming from, and figure out the best way to work with them to come to a resolution. This section discusses some situations that may arise and how to resolve them. Remember, there are two types of disagreements: those that create stronger bonds and those that cause harm, either to the team or to your customers. I focus here on the second category.

Situation: You're the Young Boss Managing an Older Worker

As a young leader, you may find yourself managing someone who's older than you. Some older employees are okay with that, while others may feel that they should be the ones doing the managing. In my research I have found that 83 percent of people have seen younger employees managing older ones.[17] Nearly half of older employees feel that younger ones lack managerial experience, an attitude that can have a negative impact on company culture. At the same time, over a third of younger employees believe that managing older employees is difficult.

When Amit Trivedi, CP infrastructure and analysis manager at Xerox, was a new manager, he had an experience with an older teammate that got him thinking differently about the generation gap. The teammate told Amit, "We have never had to change this process and it has always worked." Amit was interested in improving the process and asked a number of questions, including, "What problems would arise if the process were changed?" and "Is there any possible way to make the current process more efficient?" The resulting discussion gave Amit and his colleague a chance to explore each other's perspectives and to find a way to collaborate on ideas.

Let's imagine that one of your older direct reports is frustrated with you because you always think you're right and you never take into consideration their opinions—which are based (in their view) on a lot more experience than you have. What do you do?

First, you need to gain insight into the employee's point of view before drawing any conclusions or pushing your view on them. Second, adapt your communication style to their preferences. If they want a face-to-face meeting, don't force them to email you. Older workers are often more "traditional," and you need to speak their "language" to get on the same page and show them the respect they desire. Finally, keep an open mind and don't make assumptions about how their age might influence their thinking or abilities. Make them feel involved and valued by incorporating their ideas into your final decisions.

Everyone I spoke with agreed that most employees—regardless of age or job title—want the same thing: to improve their company and its products and services. No generation has a monopoly on the "right" way to achieve that goal. Nim De Swardt, chief next generation officer at Bacardi, put it nicely when she told me, "All generations need to understand that in this new world of work we are less bound by convention, structure, and hierarchy, and that experimentation is essential for innovation."

Situation: You're a Different Nationality Than Your Employee

As discussed, the workplace is becoming more diverse by the day, and sooner or later you're going to be working with people with whom you don't share an ethnic or geographic background. Let's say that you're Chinese and that you and an American colleague are putting together a bunch of slides for a presentation that you are to jointly deliver at an upcoming conference. Because of your differences, you find yourself spending a huge amount of time talking about how to divide the work and presentation time and dealing with a bunch of other issues that you're pretty sure would never have come up if your copresenter were more like you. What do you do?

First, sit down with your copresenter and ask them how they think the presentation should be organized and how to divide up the responsibilities. After they've given their opinions, share yours. This way you're showing respect and conveying that you care about what they have to say and the role they want to play. If there are any language barriers or you don't feel comfortable with how certain words or phrases might translate for your audience, ask them to help revise them. Finally, you should both agree on who's responsible for which slides and what the timeline is for preparing the final presentation.

Measuring Diverse Ideas

Despite the power of human connection, we're still going to need hard metrics to prove the value of our hard work. The good news is that it's possible to measure diverse ideas in many ways that provide value to your team, company, and customers. By conducting a 5-point employee satisfaction survey every quarter, you can measure how diversity is impacting employee happiness, well-being, and productivity. If you move from a 3 to a 5 on overall employee satisfaction, you're doing a great job. In the survey, ask specific questions about diverse ideas, such as, "How have you benefited from a diverse team?" and "How much has diversity impacted your overall performance?" If the answers are positive, your team could be a model for other teams or your company as a whole. You can have a major impact by influencing other teams through the decisions you make about diverse ideas.

Another way to measure diverse ideas is the number and quality of the ideas your team generates. If the team's ideas result in a successful project, it should be easy to measure the resulting cost reductions, revenue generation, or productivity increases.

Key Takeaways from Promote Diverse Ideas

1. **Recruit the right team members.** Be conscious about incorporating new voices. Change your hiring process so that the criteria are broader and go beyond education, geography, and so forth. When you hire right, diverse ideas will happen naturally.
2. **Create a safe, supportive culture so your teammates can freely share new ideas.** People are more inclined to participate and collaborate when they can lower their guard and feel comfortable.
3. **Put technology aside before getting to know your teammates' and employees' beliefs and viewpoints.** While tech can be useful in encouraging people to contribute wherever and whenever they want, a sit-down meeting will give you a far better sense of each person as an individual.

Chapter **5**

Embrace Open Collaboration

Deliver the seed of an idea and then ask for help to make it better—it doesn't need to be totally perfect the first time around.

—Beth Comstock[1]

How we collaborate and network within our teams has evolved over the past decade, and we now rely to a much greater extent on technology for both activities. Today, you can have a videoconference with teammates and coworkers in a dozen different countries without even leaving your office (or your bedroom, for that matter).

After surveying thousands of employees across all age groups globally, I've found that most people prefer in-person communication to using technology. And when I ask about their desired office environment, everyone chooses corporate office over working remotely. But despite our claims that we prefer deeper connections with our colleagues, out of habit we continue to rely way too much on technology that, while admittedly allowing us to collaborate efficiently, inadvertently weakens the very relationships we say we want.

In a recent study Pew Research found that 40 percent of younger workers spend 30 percent of their personal and work time on Facebook and choose texting, email, and video chats over in-person communication. This has created a generational divide between more tech-savvy younger workers and less tech-savvy older ones. As new tech-based communication tools come online (where they're quickly adopted by younger workers), this divide becomes wider and deeper and causes work conflicts. For example, our research shows that just over a fourth of younger workers want virtual reality integrated into the workplace. Using this new technology may seem cool to the younger worker but is quite unnecessary and counterproductive when trying to settle a dispute, especially with an older worker.

To better understand how we got ourselves into this situation, let's examine how workplace communication and collaboration have changed over the past few decades. After that we'll take a look at what we need to do to make the workplace more functional—and more human.

Connecting and Collaborating Require New Skills

Not all that long ago, when organizations were structured hierarchically, leaders at the top of the organizational chart controlled the flow of relevant information. But as technology has evolved and organizational structures have flattened, just about everyone—regardless of job title—has equal access to that information. So do customers, because smart companies have realized that giving customers more information improves their experience.

As the *what* part of communication has grown, the *when* part has grown even faster. Back in my parents' day, a traditional full-time job was 9:00 a.m. to 5:00 p.m. Today, business is 24/7, and more than half of managers expect their employees to respond to email and phone calls outside of office hours.[2] Whenever I speak to groups, I always ask my audience how many of them check their email on vacation, and I rarely see unraised hands.

And then there's the *where* part of communication. In the 1980s, employees wanted offices without distractions, so companies built cubicles for them.[3] In the 1990s, employees seemed willing to give up privacy in favor of having more direct access to and personal interactions with others, so the open office movement was born.

Today, employees are back to wanting more privacy and less noise. But while many organizations are trying to come up with yet another one-size-fits-all office design, the most successful offices are being designed with flexibility in mind and include regular offices, cubicles, conference rooms, huddle rooms (smaller conference rooms for quick meetings), lounge areas, meditation rooms, cafés, and outdoor spaces.

Thanks to technology and the ever-growing proliferation of devices, we can do our work literally anywhere on earth. As an employee, you can work at home, at a coffee shop, or even on a plane or in some other vehicle. As an employer, giving your employees a choice of where and how to work is critical to ensuring that they're comfortable, feel supported, and are in a creative environment. In our research we have found that a growing percentage of workers are willing to make less money in exchange for this flexibility. We believe that the most successful leaders of the future will embrace flexibility.[4]

The workplace continues to evolve as we experience cultural, social, demographic, and technological changes. When these changes occur, we have to account for new employee preferences, priorities, and behaviors. The following chart compares the past and present workplace.

	Past	**Present**
Structure	Hierarchical	Flattened
Work Schedule	Structured	Flexible
Information	Isolated	Shared
Dress	Business	Casual
Location	Centralized	Dispersed
Environment	Solitary	Choice
Meetings	Formal	Spontaneous

The most important part of communication is, of course, the *who*. And like the what, when, and where parts, the *who* (perhaps better known as "meetings") has evolved quite a bit over the years.

Long ago most meetings took place in person and were large, formal, and well choreographed. Today, meetings tend to be smaller, more casual, and more spontaneous. Technology, of course, has played a vital role in this evolution, enabling a geographically dispersed workforce to share ideas, collaborate, and connect in ways that never would have been possible even a decade or two ago. As technology continues to evolve, the definition of what a meeting is will continue to be challenged.

As you can tell, I'm a huge champion of technology, flexibility, and open collaboration. But I also see the downside: weaker relationships and less overall happiness. Although working remotely gives us freedom of choice, it also tends to separate us from the relationships that make businesses function properly. I've felt lonely working from home for years and have started forcing myself to go to the office and have meetings at various locations around the city. When you don't have true human interaction, you lose some of the humanity that makes collaboration meaningful, fun, and exciting. One study in the *Harvard Business Review* found that the most productive and innovative teams had leaders who were both task and relationship oriented.[5] Leaders who focus only on results will be ineffective if they disregard the relationships that are required to achieve those results. In-person relationships are much stronger than virtual ones.

Teams that collaborate solely using technology form weaker relationships, while strong relationships create more team commitment and reduce turnover. In 1977, MIT professor Thomas J. Allen studied the communication patterns among scientists and engineers and found that the further apart their desks were, the less likely they were to communicate. If they were thirty meters or more away from each other, the likelihood of regular communication was zero.[6] Face-to-face meetings give you the proximity and presence that make collaboration more effective. Mike Maxwell, KitchenAid global category leader of the culinary-global leader program at Whirlpool, said, "Using technology can feel cold, and I am less willing to ask something that may make me seem incompetent. I am also better able to read the room and pick up on the unsaid words. Reading the room is critical for knowing when things need further explaining or when to drop something that isn't going over well."

Many companies are shifting away from remote working because they feel the best ideas come from chance conversations between people at the "watercooler," which tend not to happen when employees work remotely. Some of the biggest technology companies are investing more money in their office designs to encourage interactions. Apple's headquarters in California may resemble a UFO or possibly the Pentagon, but with more than 2.8 million square feet of office space, it can accommodate about twelve thousand employees. The purpose of the enormous facility is to encourage collaboration between workers and between departments. Jonathan Ive, Apple's chief design officer, wanted to "make a building where so many people can connect and collaborate and walk and talk."[7]

In this chapter I help you better connect and form the type of deep relationships with your coworkers that you'll need to stay engaged, fulfilled, and productive. Let's start out with a quiz to help you see how much you rely on technology tools as a crutch when collaborating in the workplace.

Self-Assessment: Do You Rely Too Much on Technology to Collaborate?
This brief quiz will give you a sense of how much you rely on technology instead of picking up a phone or meeting someone in person. The higher your score, the more negative the impact of technology is on your work relationships.
The first thing I do when I wake up in the morning is check my email. 1—Never 2—Rarely 3—Sometimes 4—Very often 5—Always
I try to avoid meetings because I can email or instant message instead. 1—Never 2—Rarely 3—Sometimes 4—Very often 5—Always
I actively respond to emails when I'm not at the office instead of picking up the phone. 1—Never 2—Rarely 3—Sometimes 4—Very often 5—Always

How often during the day do I check my phone to see if I have another email or text? 1—Never 2—Rarely 3—Sometimes 4—Very often 5—Always
I use tech tools mainly because I think they're more effective in getting my message across or solving business problems. 1—Never 2—Rarely 3—Sometimes 4—Very often 5—Always
I try to avoid resolving office conflicts in person and would rather email, text, or instant message to communicate differences. 1—Never 2—Rarely 3—Sometimes 4—Very often 5—Always

If your score is below 20, chances are good that you're getting the necessary face time to build stronger relationships with your team members. If your score is over 20, like most people who have already taken this quiz, you're a technology addict and need to work on detaching yourself from it in favor of more personal forms of communication.

When to Use Technology (and When Not to) for Communicating with Your Team

To help you figure out when and how to use technology, this chart explains what to do in various situations that you'll undoubtedly encounter in the workplace.

Activity	What to Do
Setting up a meeting with a colleague	Use Google Calendar, Microsoft Outlook, or another program to schedule the meeting, but make sure it's a phone call, a videoconference, or an in-person conversation—something in which you can see (or at least hear) the people you're meeting with.
Handling a workplace conflict in which both parties are emotionally upset	Skip the technology entirely. Instead, speak directly to your colleague (or at the very least, pick up the phone) to ensure that emotions are expressed and the conflict gets resolved.

(continued)

Activity	What to Do
Letting your coworkers or manager know that you're taking a sick day	Email your team that you won't be in the office due to illness but that you'll be available to answer urgent emails if necessary.
Sharing a new idea that will benefit your team and organization	Instead of emailing your team about your idea, wait until your weekly meeting to share it. This way, your team can pick it apart, help you improve it, and potentially adopt it.

Now that you have a sense of how to improve the way you communicate, let's talk about the importance of connecting in person and how to establish a more open culture. By improving yourself, you're also setting a good example for your coworkers. Conversely, failing to improve yourself will affect your overall well-being and make it harder to be a highly effective leader.

Silos: The Enemy of Collaboration

You know collaboration isn't working when you have organizational silos. When teams and departments are in silos, they don't share information with others, which reduces operational efficiencies, depletes morale, and isolates your employees. If two teams are working on the same project without knowing about each other, you have a major collaboration problem. When teams are actively sharing their goals, progress, and results with others, you never get into that situation. Instead of competing with another team to see who can finish the project first, either you should work on it together or one team should stop working on it altogether.

Silos occur when there are power struggles and a lack of cooperation. If another team isn't open to meeting with you, or seems to be holding back critical information, collaboration isn't working.

Another sign that collaboration is broken is when you have meetings in which little or nothing gets done. Time-suck meetings frustrate employees and often cause them to look for any excuse to avoid future ones. Over a third of our time each week is spent sitting in meetings, and almost half of employees believe that many of these meetings are a waste of time.[8] When meetings don't work, collaboration fails.

Use Technology to Intentionally Drive Human Interactions

Instead of relying on technology to collaborate, use it to make more meaningful connections with your team. When it comes to communicating with others, I'm something of an introvert and have always found it easier to hide behind a computer. I made more than one thousand cold calls during a summer high school internship, but I would much rather send an email than risk having someone on the phone tell me, "I'm not interested." The same rejection over email is less harsh, and I take it much less personally than a phone call or an in-person rejection. For that reason, early in my career I preferred using email and social networks to reach out to key executives and successful businesspeople. Over time, though, I was able to convert many of these email exchanges into great business relationships by meeting on the phone or in person. I've since realized that the initial digital handshake was a great way to make an introduction, but in-person meetings have created stronger bonds that have led to many opportunities in my life.

As I discussed previously, we are spending almost a third of our personal and professional time on Facebook. In addition to Facebook, we spend about 6.3 hours a day checking our email[9] and send more than thirty text messages a day.[10] While we enjoy the instant gratification and seamlessness of these platforms, they aren't as effective as they might appear. Many of us are addicted to sending and receiving messages, but a simple face-to-face meeting can save much time, energy, and emotion. (Let's not forget Mahdi Roghanizad's study that found face-to-face requests were thirty-four times more effective than email ones.[11])

In-person communication makes it easier to build trust because people get a greater sense of who you are based on your tone and body language. If you rely on technology too much, you lose the humanity and emotion required to create trust. Without face time, you're forced to rely on others to be responsive to texts, emails, instant messaging, or social media updates that become more of a liability to your relationship than an asset.

When your teammates are more worried about what collaboration tool they want to use and less about the idea of sharing information to accomplish goals, you have a problem. These tools are valuable, but the focus should be on supporting your teammates and encouraging the sharing of perspectives and ideas. Technology shouldn't be used as a crutch or an excuse; instead, it should be a starting point to drive more human connections.

Promote a Culture of Open Communication in Your Team

After surveying thousands of employees globally, I have discovered that people generally want open, honest, and transparent communication. We'd all rather have a leader who's open and honest with us than one who's confident, ambitious, committed, or even inspirational.[12] That's because openness and honesty create the trust required to form strong relationships and teams. Here are some ways to promote a culture of open communication in your team.

1. **Make everyone on your team commit to being open and accessible.** The best way to do this is to set ground rules up front for people to agree on. For instance, if you have an office, you should have an open-door policy that encourages any team member to stop by and share an idea or issue without repercussions. Try the open communication exercise below.

Open Communication Exercise

In one of your all-hands meetings, have each team member write down one success and one struggle from the past week. The simple act of writing down something that you view as successful helps you feel grateful for the work you've done, and by writing down a struggle, you're acknowledging that you aren't perfect and that you can improve. After each team member writes down both items, go around the room, encouraging everyone to share what they have written. After each member discusses their struggle, the rest of the team members should openly discuss what to do to overcome the obstacle now and to prevent it from arising again in the future. If you repeat this exercise every week, you'll start to create a culture in which both success and failure are openly talked about. That helps solve problems more quickly and makes teammates feel more comfortable. This type of open communication also helps you and your team deal with small issues as they occur, which may keep them from becoming bigger problems later.

2. **Champion real-time feedback.** Our research shows that employees (especially younger ones) want feedback regularly and are too impatient to wait a year for a performance review. Get your team comfortable with providing much more frequent feedback. This starts with you. If one of your teammates shares a new idea in a meeting in an off-putting tone, talk to him afterward and explain that although the idea was great, there are better ways of communicating it.
3. **Share your to-do list.** Although you may think you should keep your tasks and goals to yourself, sharing them actually increases the chances that you'll accomplish them. When your coworkers know what you're working on and what your priorities are, they'll be much more likely to help you accomplish them. You can all

hold yourselves (and one another) accountable by sharing your daily or weekly to-do lists. If you're falling behind, your team can help push you through.

Sample Shared To-Do List

This is a brief example of a shared to-do list that we've used at my company to plan our executive conferences. Everyone on my team is assigned a series of tasks, and we share them with one another to ensure that we're all on the same page, that we aren't duplicating efforts, and that everyone is held accountable. We all have a common goal of making the event successful and divide the responsibilities evenly among teammates so their assignments play to their strengths. Here is an example of a team's to-do list.

Colleague 1

- Reach out to potential speakers for the event and set up calls to review their content and expectations.
- Send speaker bios, pictures, and topics to Colleague 5 for inclusion on the website.

Colleague 2

- Reach out to sponsors to notify them of the conference and available paid speaking opportunities.
- Review the agenda to ensure that it's full and that there's enough time allotted for each presentation.

Colleague 3

- Collect speaker presentations and format them using our template.
- Pull the prework together for attendees to review pre-event.

Colleague 4

- Connect with executives and invite them to attend the event.
- Monitor the P&L for the event to ensure it remains profitable.

Colleague 5

- Create the event website that showcases the agenda, speakers, location, and topics covered.
- Email our database to notify those in it of the new event, and log registrations.

Colleague 6

- Call the hotel to arrange for a room block so there's ample space for attendees.
- Keep track of the attendee list, making sure everyone is signed up.

It's extremely important to build open communication into your team culture so that members won't be afraid to express their opinions and feelings—even newly recruited members who are adjusting to your work style. To do this, never penalize teammates for being honest. Instead, reward them with respect because they want what's best for the team.

Hopefully everyone on your team will want to contribute ideas. But as a leader, that wealth of ideas can put you in the uncomfortable position of having to choose some over others—after all, you can't make everyone happy all the time. Just make sure that everyone fully understands the reasons you're choosing one option over another.

Even with Open Communication, Conflicts May Arise

Despite your best efforts, conflicts at work are bound to happen, which means every employee needs to know how to manage them before they get out of hand. CPP, The Myers-Briggs Company, estimates that employees spend almost three hours each week dealing with conflicts. While older workers may pull you aside and personally ask you to help settle a dispute, younger workers tend to try to solve their problems from a distance, usually hidden behind their devices. Unfortunately that approach usually backfires, and small disputes that could have been settled in a five-minute, in-person meeting end up blowing up over the course of a week.

Because texting and instant messaging are technically forms of communication, they give the impression of creating relationships between people. But those relationships are extremely superficial. As a result, they won't help workers fully express their emotions, understand where others are coming from, or work together on a solution to the problem. At the same time, they keep workers (especially younger ones, who are more likely to communicate via devices) from fully connecting with others on their team, which makes them feel more isolated from and less attached to the very people they're supposed to be collaborating with.

How to Reduce Office Conflicts

1. **Understand your teammates' needs, work styles, and personalities.** Think about what they best respond to so that you're showing them respect. Learn about their communication preferences based on what they've done in the past. Watch for cues in their body and verbal language to further understand how casual or sophisticated you need to be when working with them.
2. **Tailor your messaging, tone, and language to eliminate any room for misinterpretation or misunderstanding.** For instance, one younger team member might want short, succinct messages on an instant messaging platform, whereas an older, more experienced worker might want a formal, in-person discussion or an email with multiple paragraphs.
3. **Encourage and support your team regularly.** Team members don't want to get into battles with those who are actively trying to make them more successful. Helping your team members solve their problems, giving them the resources they need to perform work that they may be having trouble with, or just doing something nice

like getting them coffee will make them less apt to want to butt heads with you.

4. **Ask for help.** Too many of us are afraid to ask for support because we don't want to be perceived as incapable of doing our jobs. If you find yourself frustrated with a colleague, it's a smart move to get advice from your manager or a trusted mentor. An outside adviser may be able to help you identify potential conflicts before they arise or help you solve a problem before it escalates.
5. **Set workplace conflict guidelines to stop conflicts quickly when they happen.** Create a document that includes basic steps that your team members should follow to report a conflict, as well as steps they should take on their own to start resolving that conflict. Minor issues can be handled between coworkers, whereas major ones should be reported to management.

Unfortunately, no matter what you do to prevent conflicts, you can't control what other people do, and one of these days a conflict is going to erupt. When this happens you need to be ready with a conflict management process that will keep things running as smoothly as possible and keep tasks and projects from being delayed.

How to Resolve Office Conflict

1. Listen to what the parties involved are saying before rushing in to try to resolve the matter.
2. Let team members express their feelings, opinions, and frustrations so that you learn more about their perspectives and how to approach them.
3. Clearly define what the problem is and determine the next course of action so that you can move forward to solve it.
4. Find areas of agreement that are hiding among the disagreements.

5. Brainstorm solutions with the parties so that they have a stake in the outcome. Once one solution is chosen and agreed upon, make sure that all parties commit to moving forward and that they understand what they need to do to keep the same problem from happening again.

Conflict Resolution Exercise

Think of a conflict that you're dealing with now or that you may be dealing with in the future. Then take a piece of paper (or open a document on your computer) and divide it into two columns. On the left side, write down all the stories you're telling yourself about the person on the other side of the conflict. In the right column, write down all the facts that are observable and objective. As you review both columns, try to identify one of the stories you are telling yourself that is proven false by one or more of the facts listed. This exercise will help you remove unnecessary emotions that may prevent you from resolving the conflict and will ensure that you'll handle it in a mature way.

How to Handle Various Office Conflicts

Like it or not, over the course of your career you're going to encounter a wide variety of conflicts. To teach you how to solve some of the more common situations you may run into, here I help you analyze a couple and give you sample dialogues that will help clarify things.

Managing an Emotional Employee

One of your teammates is having a bad day because they don't feel appreciated, they aren't feeling well, or their father got laid off. Some of these situations may be out of your control, but you still have to deal with the emotional aftermath, which can affect each teammate's productivity, the team's success, and your relationships with all of them. Take the time to listen to your teammates and express sympathy or concern. Perform an act of kindness for them—something positive and unexpected—such as taking them out for lunch to talk things out.

Sample Dialogue

You: I've noticed that you're down on yourself today compared to other days. What's wrong?

Emotional employee: I don't feel like I'm appreciated as a member of this team and haven't been recognized and rewarded as I'd hoped.

You: Can you give me a specific time when you achieved something as part of our team and didn't receive recognition for it?

Emotional employee: In January, after we completed our big product rollout, I didn't feel like I received any credit for being the lead programmer in charge of creating the product.

You: I'm sorry you feel that way. We'll make sure that you get the recognition you deserve the next time. Thank you for bringing it to my attention. Although I can't give you a bonus until next quarter, I'll keep this in mind.

Managing an Older Employee

If you're a manager with an older direct report, they may already be stereotyping you as lazy, entitled, or unqualified. They could be frustrated because they think they should be managing the team instead of you, which in effect is entitlement. My best advice for handling older employees is to take time to learn about their experiences and aspirations. Make sure they know that you respect their wisdom, and create a reverse mentoring situation in which you're asking for their advice, not just giving orders. Discover what their training needs are and what motivates them so you can best serve them.

Sample Dialogue

Older employee: I'm not sure I agree with you on your approach to managing this project. I've been working here for ten years, and I find that getting everyone in the room for a kickoff meeting is more effective than just starting the project immediately.

You: I respect that you've had a lot of valuable business experience and that you've been here a lot longer than I have. In my previous company, my team was more effective working independently at first before collaborating.

Older employee: Trust me, if you have a kickoff meeting, it will save everyone a lot of time throughout the entire project management process. You will be much less stressed and successful at this company if you go my route.

You: I'm going to try your method since you've done it before successfully, and I'm open to new ways of doing things. If this doesn't work for me, I will resort to my standard process, but let's give your way a chance first.

Key Takeaways from Embrace Open Collaboration

1. **Use technology to facilitate real connections.** Avoid relying on technology as your sole means of communication.
2. **Engage in open communication.** Try to help your team feel comfortable about raising important obstacles, ideas, and questions.
3. **Become an active listener.** Better listening will help you identify your team members' needs, wants, and styles. This will help you more quickly resolve conflicts as they're occurring and may even prevent some from arising in the first place.

Chapter **6**

Reward Through Recognition

Too many managers think that people are working for them; they don't realize that they should be working for their employees.

—GARY VAYNERCHUK[1]

By the time I was a teenager I had accumulated a dozen or so soccer trophies without having won many games. I was regularly praised by my parents and teachers—even when I clearly didn't deserve it—and was told that I was special. Looking back, I can see how much my ego enjoyed the compliments and awards, but I also see that it gave me an inflated view of myself (which I hope I've gotten over at this point in my life). Many of my peers grew up in the same type of environment, constantly being praised and told how great they were. Now, as adults, instead of looking to our parents, teachers, and coaches for compliments, certificates, and trophies, we turn to our managers.

Okay, so younger workers need a lot of praise. That's pretty clear. But what's especially challenging for today's leaders is the frequency with which their employees need that praise. Your team members no longer want to wait a year for a review (and raise). They want much more regular feedback. And if you're going to be an effective leader, you're going to have to give it to them.

"Older" workers accuse younger workers of being impatient and entitled. I'll admit that to some extent, they're probably right. However, one of the driving forces behind that impatience, entitlement, and associated need for nearly constant praise is technology. Think about it: the folks on your team update their Facebook status, saying that they've just finished a big project, and boom, they spend the next day or two soaking up likes, shares, "Nice job!"s, and "You're awesome!"s from everyone they know. The same thing happens if you, the team leader, post a group congratulations message online.

People on the receiving end of electronic praise feel great right away and pretty good the next day. As a result, they feel more loyal to their team, their leader, and their employer. But gradually life goes back to normal—except for one thing: the people you praised just a few days ago need more praise—and they need it now.

A New Kind of Addiction

The combination of technology, the instant gratification it has spawned, and dopamine has created a significant addiction problem. I'm perfectly serious about this. Dopamine plays a key role in what's called the "reward pathway" of the brain. That pathway is stimulated by all sorts of things that give us pleasure—food, sex, drugs, exercise, and yes, praise. In response, the reward system uses dopamine to send a message to the rest of the brain, essentially saying, "Hey, that felt really good; let's do it again." That's how our brain convinces us to eat and have sex—two activities that are essential to perpetuating the species. (When the reward pathway gets hijacked by drugs, alcohol, or other substances, addiction occurs.)

In the case of our jobs, most of us have put two and two together and know that if we work hard and do a good job, we'll get praised—and that makes us feel excellent. The problem is that because we've become so used to being praised by everyone around us, we've become addicted to it—in much the same way we've become addicted to our devices (or in the case of some people, to coke or other drugs).

In the absence of feedback, it's sometimes hard to know whether our work matters. So, as a leader, it's your job to make sure you give it if you want a motivated and happy team. When you praise your employees, they feel a variety of positive emotions, including satisfaction, happiness, and enjoyment, and they'll work hard to keep earning your recognition. But if they don't get it—just like any addict going through withdrawal—they'll feel empty, unfulfilled, unappreciated, and unsuccessful.

"A senior leader for an initiative that I was supporting recognized me after seven months of hard work," says Vicki Ng, senior program manager of Global University at Adidas. "It reinvigorated me at a time where I felt down from the drag of work and filled my energy tank for quite some time." But as important as praise is to those of us working away in the trenches, even those at the top of the company organizational chart need praise. "As CEO you often don't find yourself in a position where your employees are recognizing your work. You're generally recognizing their work, and that's the way it should be," says Rajiv Kumar, president and chief medical officer at Virgin Pulse. "There's nothing more encouraging to me than conducting a successful board meeting and hearing afterward comments like, 'This was incredible. We wish that all our portfolio companies would have board meetings like this. You were prepared, you were clear, you were concise.' That to me was one of the greatest feelings in the world, and it's where I felt the most recognition and the most meaning."

The Power of Recognition

Personal recognition can not only make team members want to work harder for you and stay at your company longer; it can also create a lasting positive memory in their minds. Katie Lucas, senior manager of digital content at HBO, has what most people would consider a really cool job. One of her early projects at HBO was working on the relaunch of the *Game of Thrones* Viewer's Guide, the ultimate resource for the series, with a huge number of updates after each episode. In order for her to publish the new website ahead of the season 4 premiere, Katie's team had late nights, dinners at the office, and weekend work sessions. "I worked so much that the brief windows of sleep I had were populated by *Game of Thrones* dreams, which, as George RR Martin fans also know, are terrifying," she told me.

The launch was successful, and months later she learned that her team had won an Interactive Emmy Award. "I couldn't believe it. I grew up just outside of Cleveland, Ohio. In my mind, Emmys were for the Tina Feys of the world, not anyone I knew." She assumed that the executives would be the only ones to receive credit for winning this coveted award and to attend the ceremony (although she still planned to include it on her résumé). Then, out of the blue, her creative director called and asked if she would like to go to the ceremony in Los Angeles. Katie did, and she spent an amazing evening with a red carpet, Morgan Freeman, and a seat at the *Game of Thrones* table with the crew that makes the series and their five Emmy statues. "It's hard to express what that night meant to me. At first it was hard to believe that I deserved the ticket. I remarked as much to my manager at the time, who pointed out that I'd put in more hours than anyone. Today, that Emmy ticket hangs above my desk."

Although many leaders don't offer much feedback or recognition, doing so can have significant individual, team, and company-wide benefits. I had a conversation with David Novak, the cofounder and former CEO of Yum! Brands, about how he created a culture of gratitude at a billion-dollar company with more than ninety thousand employees. David would hand out personalized rubber chickens, cheese heads, or wind-up walking teeth to employees who "walk the talk." He personalized his recognition and avoided giving out generic plaques or certificates, which showed that he cared and understood what would motivate his employees—or at least make them laugh! "We built a culture of recognition with each group, leader, and brand embracing recognition in their own way around the globe," he said. "As a result, we were able to reduce team member turnover from more than 150 percent to less than 100 percent by recognizing people throughout the company."[2]

Here are a few more examples of the value of recognition:

- Employees who say that they're consistently recognized at work in ways that are meaningful to them are eleven times more likely to spend their careers with one company and seven times more likely to be completely satisfied in their jobs.[3]
- Organizations that have formal recognition programs have six times greater operating margins and employees with the highest engagement levels.[4]
- Employees who are recognized are twice as likely to be highly engaged at work.[5]

Compliments Can Be More Motivating Than Cash

Money might enable you to recruit some brilliant minds, but they won't stay with you unless they feel recognized and appreciated. As my mentor Daniel Pink says in his book *Drive*, "We leave lucrative jobs to take low-paying ones that provide a clearer sense of purpose."[6] In the short-term money may feel like a sweet reward, and it's always nice to be able to buy groceries and pay the rent, but in the long-term we crave meaning. Recognition has a more powerful effect on our lives because it makes us feel that we matter. According to Norihiro Sadato, a professor at the National Institute for Physiological Sciences in Japan, "To the brain, receiving a compliment is as much a social reward as being rewarded by money."[7] Compliments can help you alleviate stress, make you feel more confident, and motivate you to strive for excellence.

This isn't to say that no one cares about money. Of course they do. However, on average, while money provides some motivation for our teammates to do their work, recognition makes them actually *want* to do it. How we feel is more important than how much we earn. In one study almost four out of five employees reported that their most meaningful recognition was more fulfilling than rewards or gifts.[8]

In an experiment by Duke professor Dan Ariely,[9] a group of employees at a semiconductor factory at Intel in Israel received one of three messages at the start of their workweek, each promising a different reward for accomplishing all their work in a day.[10] The three emails included these incentives: a voucher for a free pizza, a cash bonus, and a rare compliment from their manager. A control group received no message at all. After the first day the pizza group's productivity increased by 6.7 percent over the control group, the compliment group boosted its productivity by 6.6 percent, and the cash bonus group increased its productivity by 4.9 percent. After the second day those with the cash bonus actually performed 13.2 percent worse than the control group, and by the end of the week the cash group had lost 6.5 percent in productivity and cost the company more. At the end of the experiment, the group that received the rare managerial compliment did best of all. Clearly, personal recognition matters—even more than pizza!

A recent study by McKinsey reached similar conclusions, finding that nonfinancial incentives such as praise from an immediate manager, attention from leaders, and opportunities to lead projects were more effective in boosting engagement than financial incentives like pay, cash bonuses, and stock options.[11] Sixty-seven percent of executives, managers, and employees said praise from management was very or extremely effective, compared to 52 percent who said the same of increases in base pay. In the Virgin Pulse study, when we asked employees what would make them more engaged at work, over a third said "more recognition."[12] Researchers in Canada asked the same question and got an even stronger answer: 58 percent said "give recognition."[13]

One explanation for money's lack of effectiveness as a motivator may be that too many managers give raises and bonuses only to prevent employees from leaving, not to reward performance. Frequent, performance-based bonuses might be more effective, especially considering how much it costs to replace an employee in terms of dollars, resources, and training time.

Another problem with using money as a motivator is that although employees may receive a jolt of excitement when they first get a cash bonus, once the money's spent the excitement disappears, and they want more money. The more cash bonuses you give, the more your employees will expect. You'll never be able to keep up. It's yet another type of addiction. So before it's too late, start giving more recognition and stop relying on money to drive performance.

Throughout your day there are countless opportunities to compliment your teammates, and that can make all the difference in their work experience and satisfaction. The following chart shows five of the most common situations in which it's appropriate to compliment your teammates.

Five Situations in Which to Give a Compliment	
Situation	**How to Handle It**
In the moment	You don't always need to plan when you're going to compliment your employees so don't be afraid to be spontaneous. When you recognize someone in the moment, it comes off as more authentic.
During a meeting	Before your meeting, think about your remarks and how and when you'll deliver them. I recommend that you start the compliment by acknowledging the entire team. Then single out the individual and explain why they have made a valuable contribution. This way, you're less likely to upset the rest of the team.
In passing	If you happen to be walking past an employee you were planning to compliment, pull them aside and do it. But make it quick (and sincere, of course) because you're both probably rushing somewhere important.

(continued)

Situation	How to Handle It
Virtually	If you want to praise a remote employee, either call or set up a videoconference. That's much more personal than sending a text or emailing.
In a formal review	The easiest time to give compliments is during a formal performance review. But don't just stop at the compliment—spend some time explaining *why* the employee deserves it.

Ask Your Team Members How (and When) They'd Like to Be Recognized

As I've discussed, there are various ways to recognize your employees. Obviously not all of them will resonate with everyone. Some people, such as Katie Vachon, merchandise manager of women's apparel at Puma, will prefer public recognition in front of their peers (or a larger group), whereas others, like Chris Gumiela, vice president of marketing and advertising at MGM National Harbor, prefer a private pat on the back. "I don't need hyperbole or to be singled out, instead it is a simple note of thanks and gratitude that goes furthest."

Some may be motivated by cash, others by less tangible means. "There is no skirting the truth. I do find satisfaction from recognition. Probably more satisfaction from someone recognizing my hard work than I do from being offered a higher salary or bonus," says Stephanie Bixler, vice president of technology at Scholastic. "The lead of our consulting practice was proud of our efforts and recognized our hard work on several occasions in our team room. This commendation and recognition made me feel like my presence on earth made a difference."

The best way to figure out how to maximize the effectiveness of your recognition—and which approach/approaches will work best—is to simply ask. But be prepared for some pretty honest answers, such as the one I got from Sam Howe, director of business development at MSLGROUP. "I think there is a tendency among older generations to treat millennials like we are kindergarteners, and they offer us ice cream socials and pizza for our hard work," he said. "Now, I am for ice cream and pizza as much as the next guy, but I would like to be recognized at work, as often as earned, by being given more advanced titles (and associated compensation), responsibilities and influence. In essence, we should be rewarded by being given a voice and seat at the proverbial leadership table. And by being treated like adults, not a Trophy Generation to be placated."

Don't be afraid to have a little fun. Jill Zakrzewski, customer experience manager at Verizon, told me about taking her team on an expedition to a toy store before a recent leadership meeting. "Instead of giving out trophies or certificates, we recognized our employees with a Rubik's cube for best organized, temporary tattoos for most committed, a Transformer for most adaptable, and an Olaf figurine for most likely to keep their cool," she says. "Not only was this a hilarious award ceremony, but everyone was pleased to re-gift their awards to the children in their families."

Another big issue to think about when it comes to recognizing your employees is how often to do it. Although there's no magic number, I can almost guarantee that it needs to be more frequent than you're doing it right now. According to a study by WorldatWork, the most common type of recognition companies have is for years of service,[14] and in my experience, that tends to come in five-year increments. That may work for older employees. (Those aged fifty-five to sixty-four tend to have an average tenure on the job of more than ten years.[15]) But in the case of younger workers, who typically stay no more than three years with any one employer, most will have moved on before you've had a chance to celebrate their achievements.[16]

Some younger workers, like Laura Petti, a line producer at CNBC's *Closing Bell*, get—and appreciate—daily feedback. "After our show each day we do a postmortem meeting and discuss the highs and lows of the show," she says. "This daily feedback keeps pushing us forward, so we never settle for anything less than the best we can do each and every day." For others, once a week is plenty (although that may seem like a lot to you). Sam Worobec, director of training at Chipotle Mexican Grill, says that pretty much any interval is okay, as long as it's less than a year. "This could be once a quarter or every six months. It doesn't matter. It just needs to be more frequently than the performance review cycle," he says. "No employee should ever go to a performance review with questions about how they'll be rated, especially your senior leaders."

As you're thinking about how, and how often, to recognize your employees, there's one other important element to keep in mind: although not all employees appreciate public recognition, a lot of younger workers find it motivating. When you praise a teammate in front of everyone else, it makes you both look good. Four out of every five employees say that seeing someone else's achievements be recognized makes them want to work harder as well, in part because they want to receive the same kind of recognition.[17]

The spectator effect is especially strong among people who have been instrumental in the praised employee's success. "While being recognized for my own achievements can be rewarding, seeing my teammates win and receive accolades provides me with a greater sense of accomplishment," says Bryan Taylor, vice president and general manager at Enterprise. "When I see those who have worked relentlessly achieve their goals and I have been able to help them get there by overcoming challenges along the way, it is the most fulfilling recognition."

Finally, don't underestimate the value of a personal touch. Because we spend so much time online, it's super easy to shoot off a tweet, post something to Facebook or the company website or newsletter, or send an email blast. And while that type of praise is nice, in-person, face-to-face interaction is far more effective.

Quickly Resolve Recognition Missteps

Despite its wonderful benefits to those who receive it, recognition and praise may inadvertently lead to jealousy or other negative feelings in those who hear or see others being recognized but aren't getting what (in their minds) they deserve. They may feel bitter or angry that their work isn't being appreciated or think that the person you praised doesn't deserve it. And if they haven't received recognition for an especially long time (again, in their minds), they may feel insecure and start to convince themselves that they'll never earn your respect or praise or get a promotion. At this point, if they quit, you and your team will be shorthanded, and you'll have to figure out how to replace them. If they don't quit, I can pretty much guarantee that their productivity will suffer and that they'll start to poison the attitudes of everyone else on the team.

Here are two examples of situations in which recognizing others can have unintended consequences, and how to best handle them.

Situation 1: You Recognize One Employee and Another Is Jealous

Say you compliment one of your employees in front of the team, and another one feels that they deserve the same recognition. If you hear through the office grapevine (or from them directly) that that employee is displeased, pull them aside and explain why you recognized the other employee but not them. If, for example, the praised employee achieved a major milestone, explain to the disgruntled one how achieving that milestone had an impact on the whole company. In this way you're communicating that the employee's life was improved by the other one's success, even if the benefit was indirect. Next, set expectations by telling them specifically what you feel is worthy of recognition. Finally, choose one of the projects this employee is working on and explain that if they do it well, they, too, will receive recognition. Again, be clear about what "success" looks like, and ask what you can do to help your employee get there.

Situation 2: You Recognize an Employee Who Didn't Deserve It

You may feel that one of your employees is unhappy or isn't getting much love and want to do something about it so they don't quit. The next time you see that employee, you'll probably give them a compliment, intended to be a pick-me-up. But what if another employee—one who gets a lot of praise from you—overhears what you said to their teammate and becomes frustrated and confused. They've been a top performer, and in their view the other person is either lazy or lacks the necessary skills, or both. To keep this situation from exploding and causing problems among the team, you need to help them understand your intent so they don't feel left out or unappreciated, or think that you're favoring an underperforming employee.

The next time, don't compliment without being genuine; if the person you compliment turns out to be incompetent and needs to be fired, you'll have a tough time explaining to your legal department why you just complimented them.

Recognize Team Performance

When it comes to giving compliments, why limit yourself to individual employees when you can (and should) recognize your entire team? That way, people won't feel left out or jealous. Recognizing your team as a whole also creates a sense of belonging and reinforces relationships between employees. You can recognize the team weekly, monthly, or on a project basis. Here's how.

1. Share a specific story of how the team accomplished a goal by pointing out each team member's contribution and talking about the specific behaviors, strengths, and outcomes that made the goal a reality. For instance, if

someone on your team prepared a sales presentation and another closed a new account, point out both accomplishments. Without the presentation, the salesperson would have been unprepared to close the deal, and without the salesperson, the presentation would have just sat there and never been translated into new business. If you happen to have a slacker on your team, pull them aside and have a talk about what needs to change. Give clear, specific feedback, achievable goals, and a deadline by when you want to see measurable improvement. If the employee doesn't comply, it may be time to start looking for a replacement.

2. Encourage peer-to-peer recognition. Compliments from you, their leader, are great, but team members need to know that the people they work with appreciate them. Set aside some time at the beginning or the end of a meeting for each team member to say something she appreciates about another. Push for 360 evaluations, in which teammates offer feedback to one another, so that they get into the habit of not just improving themselves but also contributing to the success of others.
3. Evaluate your team's performance. In addition to the individual evaluations you no doubt do, start doing regular team reviews, noting the overall accomplishments and areas that need improvement. Every quarter (or six months or a year or whenever works best for you), hold a meeting to discuss where you are and how you got there, where you're going and how you're going to get there, and what your team needs to do to be even more effective than they already are. Ideally, this won't be a just-you type of review. You want all team members to participate, share their progress, and talk about the obstacles they're facing. This will help your teammates better support those who are struggling.

Create a Culture of Gratitude

As a leader, in addition to praising your employees, it's important to express your gratitude to them. The difference between praise and gratitude is subtle but important. In a few words, praise is about making them feel good about themselves, whereas gratitude is about expressing how appreciative you are of their efforts. Put a little differently, praise is an attaboy, whereas gratitude is a sincere thank-you. Again, the difference is subtle but important. Although about half of us regularly say thank you to people we're related to, only 15 percent say thank you at work, and more than a third of those in a recent survey reported that their manager *never* says thank you.[18] Another study found that the workplace is one of the least thankful places around, with 60 percent of people never expressing thanks at work at all.[19] Amy Linda, manager of global talent at Estée Lauder, sums it up nicely. "It takes so little for someone to show appreciation, but people rarely do it." Now that's depressing, isn't it?

The big irony is that showing gratitude is an incredibly effective (and equally low-cost) way of increasing productivity. In one study a university divided its fund-raisers into two separate groups. The first contacted alumni the same way they always had: by picking up the phone, making calls, and asking for money. The second group received a motivating talk from the director of annual giving, expressing appreciation for the work they were about to do. A week later the second group had made 50 percent more calls than the first group.[20]

Being on the receiving end of sincere gratitude can be quite motivating. But so can being on the giving end. Gratitude is such a powerful force that it will prevent you—and your team—from getting into arguments that are counterproductive and lower morale. If you're grateful for your teammates' hard work and support, you'll be able to prevent arguments before they occur. According to research by the University of Kentucky, participants who ranked higher on gratitude were less likely to retaliate against others even when given negative feedback and were more empathetic and less vengeful.[21] Aside from preventing work conflicts, gratitude can reduce social comparisons and unnecessary competition, because appreciation brings people closer together rather than pushing them further apart.

The simple act of regularly acknowledging what you're grateful for is so powerful that it can produce significant results even if you don't tell anyone else about it. According to Harvard Business School professor Francesca Gino, people who count their own blessings are more attentive, alert, energetic, happy about life in general, and apt to engage in health-promoting behaviors, such as going to the gym.[22] Other studies have found that grateful people sleep better, are less likely to get sick, have lower blood pressure, feel more connected to others, and are more helpful.

Creating a culture of gratitude is truly a win-win for everyone involved. If you tell your employees what you're grateful for and create a workplace environment in which everyone feels supported, they'll spread positive feelings to others by helping them on projects or recognizing their good work. In other words, gratitude is contagious—people who receive it are more likely to "infect" others. Grateful people manage stress better and are less likely to feel destructive emotions such as envy or resentment. They're also happier with their jobs. And as you no doubt know, happy employees work harder and make for happier customers. And happier customers lead to happy shareholders.

The following is an exercise that will give you some insight into how much gratitude you're showing to those around you.

Individual Exercise: What Are You Grateful For?

Over the next hour or so, jot down everything that pops into your mind that you're grateful for in and out of the workplace. For instance, as an introvert who performs best working at home, I'm grateful for being able to work with business partners who understand and appreciate my need to work remotely full-time.

I'm also grateful for my parents, who support me despite all the failures and mistakes I've made in my career. I am reminded of their generosity and patience almost daily as I confront big obstacles and challenges. Try to come up with at least three personal and three work-related statements of gratitude.

After you've finished this individual exercise, schedule a meeting with your entire team and do the following exercise. This way, you are reflecting on yourself before you hear your team's thoughts about what they're grateful for in those they work with. If you don't know what you're grateful for, how are you supposed to appreciate the contributions of others? I believe that you first have to *become* a grateful person before you start to encourage gratitude in others. Practice what you preach!

Team Exercise: Being Grateful to Others

In 2016, I went on a journey to Israel with REALITY, an initiative of the Charles and Lynn Schusterman family foundation, with forty-nine other storytellers, including film directors, journalists, and even one of the original cast members of *Hamilton* on Broadway. On the first day of the trip we broke into smaller, more intimate teams, and on the last day we all took a piece of paper, wrote our name at the top, and passed it to the right. When you received the paper from the person to your left, you'd have to write something nice about the person whose name was at the top. For instance, for one of the group members who had a positive impact on my experience, I wrote, "Thank-you so much for supporting me during this trip, it meant a lot and made it a more meaningful experience for me." When finished, we passed our papers and repeated the process until we'd had the chance to write our feelings about each person.

Set up an hour-long team meeting and bring pens and paper to distribute to everyone. You can do this in a conference room, someplace outside your office, or anywhere else that's private and where your employees will feel comfortable sharing how they truly feel about one another. This exercise will make everyone, including you, feel a deeper sense of connection and gratitude. Ideally, your team members will feel comfortable enough with one another to be okay with reading their thoughts out loud. But if that isn't the case, ease into this.

Showing gratitude doesn't take much effort, yet it can make a significant impact in your work relationships and overall well-being. Here are three ways to show gratitude to your teammates.

1. Instead of texting or emailing a compliment to one of your employees, acknowledge them in person in front of your team. This will encourage your teammates to work harder to receive a similar compliment and show that hard work is valued in your organization.
2. Make a small gesture that surprises and delights a teammate. For instance, instead of messaging that you appreciate their hard work, write them a note on a piece of paper. This will show more effort and care than an electronic message. You could also take them out for lunch or put a gift on their desk, such as a mug or a gift certificate to their favorite restaurant.
3. Be specific when showing your appreciation. Give a concrete example of something your teammate did for you and how it made a difference in your life. For instance, talk about a time when they helped a colleague solve a problem and how it assisted the team in accomplishing its goals. This way, the teammate can quickly reflect back on the work they did and understand why it mattered to the team.

Key Takeaways from Reward Through Recognition

1. **Practice the art of gratitude.** First, begin by realizing what you are grateful for. Second, tell your teammates how grateful you are for their efforts. When you show your gratitude by going above and beyond for your team and then tell them how

grateful you are, they'll start doing the same. The outcome is a team in which everyone appreciates everyone else, and that contributes to a healthy culture for everyone.

2. **Be conscious about whom you're complimenting, and why.** Be aware that your actions can create jealousy and resentment among the team. Be as genuine and honest as possible, and if you find yourself complimenting someone for the wrong reasons (for example, because you're afraid they might quit), set up a time to have the important discussion to clear the air.
3. **Recognize team performance in addition to individual performance.** This will build a stronger culture and help team members strengthen their work relationships and appreciation for one another. Business is a team sport, and you need the combined contributions of every employee to meet all your objectives. By reinforcing the power of the team instead of focusing only on the individual, you help create the synergy and connection required for long-term success.

Part III

Build Organizational Connection

Chapter 7

Hire for Personality

Personality before CV. A person who has multiple degrees in your field isn't always better than someone with broad experience and a wonderful personality.

—RICHARD BRANSON[1]

Over the past decade, hiring has changed for the better—and for the worse. But one thing that has remained constant is that when you make the right hiring decisions, you advance your team, your company, and your own career. However, because the pace of business is constantly speeding up and companies are always looking for ways to save money, many have looked to technology to lower the cost of recruiting talent and increase the number of people they can reach.

While these companies tout how much money they're saving by doing their interviewing by phone or video, they don't seem to realize that neither of those approaches can ever replace in-person interviews, in which you actually meet people, see their body language, and observe how they handle themselves. In short, those approaches are missing the critical emotional connections and personality traits that will help you hire the best possible candidate, who will stay with you longer. This is huge. Hiring someone who doesn't fit with your company's culture or can't work with the rest of the team will have a measurable negative impact on your ability to compete, keep customers happy, and adapt to change. It can also send a ripple effect through your team and cause others to question their overall commitment.

Some job seekers believe that technology has made the interview process more efficient, but most feel that it causes frustration, lacks transparency, is less personal, and doesn't provide the essential feedback they seek.[2] They fare much better when online assessments and automation bring them closer to a personal interviewer than when that technology removes the humanity from the experience. Job seekers benefit from technologies that help them find jobs but require human interaction to make the right employment decision. Whom you work with is just as important as, if not more so than, where you work or what you do.

The bottom line is that any cost savings from using technology in the hiring process are more than offset by additional expenses and other losses associated with hiring the wrong candidate.

Hiring the Wrong Employee Has Consequences

The smaller your company, the more pain hiring the wrong employee will cause your team and your company. If you're leading a start-up and the second employee you hire doesn't work out, that setback could be serious enough to make your company fail. Tony Hsieh, the CEO of Zappos, has said that bad hires have cost his company more than $100 million.[3] One study found that the cost of a bad hire is two to three times the employee's salary,[4] and when we interviewed a few hundred employers with Beyond.com, we found that it costs about $20,000 to replace entry-level workers.[5] Other studies have put the direct costs of a bad hire at between $25,000 and $50,000 per employee.[6] Jennifer Fleiss, cofounder and head of business development at Rent the Runway, agrees. "Hiring right is the most important key to productivity—in particular because when you get it wrong it takes up a ton of time," she says.

The following chart lists some of the biggest direct and indirect costs (which can often be far more significant than the direct ones) associated with hiring the wrong person.

Direct Costs	Indirect Costs
Recruiting	Increased stress on current employees during the three to seven weeks it takes to hire a fully productive worker
Job advertising	Productivity loss
Interviewing	Decreased morale
Training	Knowledge loss
Severance	Reduced work quality
Background checks	Decreased customer satisfaction
Onboarding	Damage to your company's reputation
Lawsuits	

When I was just starting out in my career, the head of communications told me the story of one employee who was incredibly talented yet toxic to his team. He continually showed up late, complained, spread rumors about his coworkers, and generally had a lousy attitude. They didn't immediately fire him because of his superior work, but his teammates ended up leaving, and eventually the company laid him off. The point is that just because someone is talented doesn't mean they're the right person to hire. A talented-yet-toxic employee will end up costing you more than they're worth.

Technology Versus Humanity

Because relationships are the cornerstone to a healthy workplace, shouldn't we put more emphasis on personality when recruiting new employees? It's challenging to work with someone we don't like, but it's exciting to work with someone who has a great personality that meshes well with our own. Hard skills are important, but they can be learned on the job. It's the soft, intangible skills that are so valuable to creating a team that thrives. They're also the ones that technology has a difficult time assessing.

As companies experiment with using machines, predictive algorithms, bots, and artificial intelligence to do their recruiting, we need to take a step back and really think about our objective. Recruiting, at its core, should be focused on matching the right talent with the right job and team. As we continue to invest more in machines, we lose track of the actual connections that make for good hires and work friendships. Companies are using machines to eliminate bias, assess human qualities like personality, scrub résumés to identify and analyze word choices, and scrub social media posts to review gestures and emotions. Although this might help narrow down hundreds or even thousands of applicants, at the end of the day only a human should be making hiring decisions, and we can't rely on these tools to make those decisions for us. In our Virgin Pulse study, 93 percent of people agreed with that thought.[7] But what worries me is the remaining 7 percent—and my suspicion that humans are gradually being removed from the hiring process by a growing number of tech-based options.[8]

To start with, using technology to interview is rife with complications. For example, candidates must have a good Internet connection, which, shockingly, isn't always guaranteed. I once lost reception during a job interview and immediately got rejected for the position even though I was qualified. Poor connectivity can also cause delays, which can make candidates seem less competent or give rise to misunderstandings. (How many times have you been watching the news and wondered why a reporter in another country is goofily nodding and not responding to a question posed by the US-based anchor?) Few candidates will have their home set up with the lighting, sound, backgrounds, and makeup artists that are optimal for the perfect interview. And let's not forget about the introverted candidates and others who are camera shy and might not perform as well in a video as they would in person. All in all, while using tech might be easier, it's simply not a pleasant way to be recruited, and it's a horrible way to make a final decision about a candidate.

A perfect technological connection is no guarantee, either. Sam Worobec, director of training at Chipotle Mexican Grill, explained to me why he'll never use video for another candidate interview. "I hired someone that was a rock star on camera. He gave all of the right answers, had a great personality, and had the demo reel to show how talented he was. During the face-to-face interview, we breezed through the process, and the whole team fell in love with the guy. Two weeks later I let him go," says Sam. "He was completely self-serving and arrogant. He was extremely charismatic, but he was a cancer to the team." His advice? "Digging in during a face-to-face interview is the only way to really get to know if someone will be a great fit for your team. The video interview is just a weeding process."

Mike Schneller, associate director of talent acquisition for Biogen, agrees with Sam. "Throughout the course of an in-person interview, you are given the opportunity to understand the person in front of you for who they truly are; there is no technology for them to hide behind, no cell phones, no video conferencing, no email. It is just you and the candidate, discussing what could potentially be a life-altering decision for the both of you." Mike believes that you can make the wrong hiring decision when technology is present, because people have a false sense of confidence, which can keep you from seeing what you really need to see: their honesty. "Human-to-human interaction is the only honest connection we have left during the interview process; let's not overlook the value of a handshake," he told me.

It's also important to keep in mind that job interviews are a two-way street. Sure, the candidates need to impress you. But you have to impress them as well. When you have in-person interviews, you're giving them an important opportunity to meet you, observe the office environment, get a taste of the corporate culture, and get to know some of their prospective teammates.

Technology Versus Likability

One of the biggest challenges with high-tech recruiting and screening tools is assessing one of the most important interpersonal intangibles of all: likability.

In one study, professors at McMaster University in Ontario, Canada, found that applicants who are interviewed through video come across as less likable and are less likely than those interviewed in person to be recommended for hiring. At the same time, candidates rated the interviewers as less attractive, personable, trustworthy, and competent.[9] Other studies have found much the same thing,[10] and the conclusions are inescapable (although far too many companies aren't making the connection): (a) using technology to interview is bad for both the interviewer and the interviewee; and (b) while tech-based interviews may serve a valuable purpose, if you need to do initial screenings of large numbers of candidates, when it comes to the final hiring decision, in-person interviews are essential.

Om Marwah, global head of behavioral science at Walmart, reached similar conclusions based on his own leadership experience. He believes that we lose out on nonverbal communication when we're on video calls or the phone. The capacity to feel empathy and absorb ideas is magnified in person, so when you're not, you're only getting a slice of the quality of conversation that could take place. "I'll often fly across the country just to do an in-person meeting like many other people simply because of these reasons," says Om.

Hire for Personality to Promote a Positive Work Culture

A lot of people are smart enough and have the skills to do the job you're hiring for, but not as many have that unique combination of personality traits that will make them a good fit with the rest of your team. And because skills can be taught and personality can't, I always recommend that managers hire for personality and fit, then train for skill. If you don't get along with an employee or if she doesn't have the right attitude or work ethic, it will negatively impact your entire team. Many—but far from all—companies already realize this. In a study we did with Beyond.com, employers told us that cultural fit is the single most important hiring criterion, more than experience, coursework, GPA, and education. And the top three skills employers are looking for are soft: positive attitude, communication, and teamwork.[11]

The following are top CEOs' reasons for hiring for personality.

Top CEOs Hire for Personality	
Robert Chavez, CEO of Hermes US	*When it comes to hiring, we look for people who have a sense of humor, people that can smile.*[12]
Elon Musk, CEO of Tesla and SpaceX	*My biggest mistake is probably weighing too much on someone's talent and not someone's personality. I think it matters whether someone has a good heart.*[13]
Howard Schulz, chairman and CEO of Starbucks	*Hiring people is an art, not a science, and resumes can't tell you whether someone will fit into a company's culture.*[14]

Cultural Fit Versus Diverse Ideas

Lauren Rivera, a professor at the Kellogg School of Management at Northwestern University, studied the top professional service firms, looking at the impact of cultural similarities between job seekers and employers.[15] She found, not surprisingly, that employers seek candidates who are similar to them in terms of culture, experiences, goals, and work styles.

In chapter 4, I discussed diverse ideas and why it's important to deliberately create a team that includes members who have a variety of backgrounds, experiences, cultures, and so forth. What I'm saying here about cultural fit in no way contradicts what I said about diverse ideas, and I encourage you to build the most diverse team you possibly can. That said, workplace cultural fit is a little harder to define than overall culture. The issue here is how people will mesh. The woman who shows up for an interview with a conservative investment bank wearing jeans and a T-shirt might not be a good fit. And neither would the shy young man who's applying for a high-powered sales job but who has trouble making eye contact. You wouldn't hire a clothing designer who wasn't detail oriented, an obese personal trainer, or an anxious surgeon.

Carly Charlson, senior manager of public relations at Best Buy, sums up the issue of fit versus diversity nicely. "When hiring, I look for someone who has the same values as me. This doesn't mean they need to be like me—different backgrounds, styles, experiences can be a huge asset to the team," she says. "But finding someone who shares my core values at work: open communication, a willingness to learn, and a team-first attitude, make all the difference." Besides asking about work-related issues, Carly asks candidates about what they like to do in their free time. "It sounds like a throw-away question, but I'm shocked how frequently it says something deeper about who they are, what they like, and how they'd fit in with the rest of our team."

Five Personality Traits to Look for in a New Hire

When you're hiring for your team, there are five personality traits that you should look for. You'll be able to screen for each by paying close attention to the answers you get to a number of key, strategic questions. Let's take a closer look at each of these traits, how it makes for a good employee, and how to determine whether or not the candidate has it.

Confidence

Many of the employers and hiring managers I've spoken with tell me that confidence is a trait that too many applicants are missing. If you're not confident, you're less likely to share new ideas, stand up for what you believe in, and perform at your best. You second-guess yourself and come across as less competent than you actually are. When you're confident, you know what you're doing and how to convey your knowledge to others. When one person lacks confidence, it permeates through the rest of the team and hurts everyone's overall performance. As the interviewer, you will naturally lose interest in someone who doesn't make eye contact, has a weak handshake, doesn't speak clearly and coherently, or "up-talks" (makes statements that sound like questions). These are often signs of people who lack confidence or aren't sure what they're speaking about. Confident employees are more effective at teaching and helping others and at getting work accomplished without stress and distraction. Sometimes you'll interview someone who appears shy or has a stutter, which could convey a lack of confidence. Instead of immediately judging them, assess their subject matter expertise, the questions they ask, and how they dress. Try to make them feel comfortable and safe, which will make them more likely to be open with you.

Interview question: What was an obstacle you overcame in a previous job?

The answer to this question lets you know about the candidate's ability to push through failure and the challenges that naturally happen in any work situation. An unconfident employee wouldn't have the tenacity to find a different solution to a problem, whereas a confident one would have found a way through.

Attitude

You want to hire employees who have a positive attitude because they tend to boost the morale of everyone around them and encourage and motivate their team members to perform better. Conversely, employees with a negative attitude are generally horrible to be around. They can detract from the whole team's performance and the overall culture of the company, and they often cause others to want to switch teams or leave entirely. "Of course, you always hope to find the smartest and most talented people," Simon Bouchez, CEO of Multiposting, told me. "But I quickly realized that someone bringing a positive attitude toward your product, your business, and your team creates much more added value than any smart but less enthusiastic person."

Consultant Mark Murphy tracked twenty thousand new hires, looking at the impact of attitude on career trajectory. Mark told me that he found that when new hires failed, 89 percent of the time it was for attitude reasons and only 11 percent for lack of skill. Those who had a negative attitude were harder to coach and had lower levels of emotional intelligence, motivation, and cultural fit. Basically, the negative attitude got in the way of their ability to produce superior work, get along with their colleagues, and be satisfied at their company.[16] When inevitable changes and challenges happen at work, teammates with a positive attitude can more easily cope with them and remain levelheaded instead of panicking.

Interview question: When have you admitted to your teammate(s) that you made a mistake, and how did you manage it?

This question is designed to get candidates to tell you how they handle mishaps at work. Those with a positive attitude will usually tell you that they apologized to their team and will explain how they'd handle things better the next time around. Those with a negative attitude will generally talk trash about their former team or use a tone of voice or body language that conveys the same message: it's everyone else's fault, not theirs. Candidates with a positive attitude will hold themselves accountable and tend not to make excuses or point a finger at others.

Professionalism

The most obvious signs of professionalism are punctuality at the interview (or being early) and the candidate's basic manners. The second candidates enter the room, you can make a quick judgment about the impression they're making and whether they'll be a good fit for your organization. The way they dress has a major impact on how they feel about themselves and how they'll perform. In one fascinating study,[17] subjects were asked to switch between formal and casual clothing before taking cognitive tests. Those wearing formal attire were more creative and better able to solve problems.

When you wear formal clothing, you feel more powerful and in control, and you take yourself more seriously. When a candidate comes in for an interview late, wearing a T-shirt, he clearly isn't taking the interview seriously, and you shouldn't, either.

Interview question: Give me an example of a situation in which you had a conflict with a team member, and tell how you handled yourself.

Listen carefully. The answer to this question will give you a sense of how candidates handle their emotions during a tough situation. Sooner or later, an on-the-job conflict will arise, and you want employees and team members who can carry themselves professionally and do everything possible to resolve the conflict in a way that doesn't damage the team.

Likability

I have two friends who definitely have the likability factor. They exude positive energy, and they're a pleasure to be with. In the workplace, likable people somehow manage to bring out the best version of you. Also, they have an incredible competitive advantage because they tend to get promoted more quickly (managers tend to promote people they like over those they don't) and build strong relationships with others, which leads to new opportunities. Since 1960, Gallup has published a personality factor poll prior to every US election, and these polls show that likability is one of the three consistent factors that predict the final election result. People can't relate to candidates they don't like and won't vote for them. In a study published by the American Psychological Association, researchers compared likability and self-promotion.[18] Likable candidates were perceived to be a better fit and were more likely to be recommended by a recruiter or hired directly for the job. Candidates who were big on self-promotion, on the other hand, either had a neutral or a negative result on the hiring decision.

Interview question: Who has been a great mentor to you, and how was that manifested?

The answer to this question will give you some clues about candidates' relationships with others. Likable candidates typically attract better mentors and describe those relationships in a more positive way. Someone who claims not to have any mentors may be arrogant or a know-it-all, or just hasn't invested the time in seeking out support. Everyone needs a mentor!

Curiosity

You want candidates who are curious about your background, your executives, your products, your company, and your industry. People who are curious about their own potential and are willing to try new tasks and roles are better able to adapt to change, challenge themselves, and grow as team members. This trait is defined by problem-solving abilities, an ongoing need to learn, and a strong commitment to a team. Curious people are more likely to want to learn from others and are open to team diversity. One study found that 57 percent of companies are looking for candidates with intellectual curiosity.[19] Multiposting's Simon Bouchez told me that he often comes right out and asks candidates whether they've visited the company website and what they'd change.

Interview question: Do you have any questions about the position or the company?

Candidates' answers to this question will give you a glimpse into how well prepared they are for the interview and how interested they are in you and your company. If they have no questions (which happens more often than you might think), they may not have done much research on the company, products, or you, despite the fact that most of that info is no further away than their phone. Candidates should ask about the company's vision, your background, the product road map, the company culture, and what their daily schedule might look like if they start working for you. Curious people ask a lot of questions, and an interview is a two-way street. Candidates need to impress you, just as you need to impress them. People who don't ask good questions—or any at all—won't be the type of employees you want because they won't push the boundaries or challenge the status quo.

Sample Interview Questions to Help You Hire for Personality

Question	What It Tells You About the Candidate
How would you describe yourself?	This will challenge candidates to reflect on how they view themselves.
How have you handled previous conflicts with coworkers or clients?	You'll gain insight into how candidates deal with difficult situations.
How would your best friend describe you?	This will get candidates to talk about how they treat others—especially those they really care about.
How are you involved in the community?	This will give you a sense of what candidates care about and what their life outside of work is like.
How would you handle a task that seemed impossible at first?	This will help you better understand how candidates can solve problems even if the issues are challenging.
Do you prefer to work in a team or on your own? Why?	You will get a sense of the candidates' work habits and if they are capable of successfully collaborating in your team.
When have you been most satisfied in your life?	This will give you an idea of how you can best support the candidates if you choose to hire them.

(continued)

Question	What It Tells You About the Candidate
What things do you *not* like to do on the job?	You will learn about what the candidates are not willing to do and what they may believe is beneath them.
When were you excited about your work?	This reveals what motivates your candidates.
What types of activities or hobbies do you enjoy outside of work?	Aside from their community involvement, this question gives you a sense of where the candidates invest their time.

Signs You Shouldn't Hire Someone

Chances are, one of these days you'll find yourself interviewing a candidate who does something that makes you realize that they aren't going to be a good fit for your team or your company. It might be lack of eye contact, a weak handshake, showing up late, laughing too much, or something else. Or it could be a lack of passion about working with you and your organization. Other red flags include energy level, attitude, and even the number of questions someone asks during the interview. When you ask candidates about a time when they felt they had a successful work experience or excelled on a project and they can't pinpoint one, that says a lot—and none of it's good. Typically, they lack self-esteem or haven't had enough experience or results in their prior work history.

Pay close attention if you get a sense that a candidate is exaggerating their achievements. If what they say is far removed from what's on their résumé or online profile or what you know is true, be wary. They may be embellishing because they're nervous or want to impress you. Or they're demonstrating that they're dishonest or not a team player. People who take credit for their team's hard work are often selfish, and if you hire them, they may not be able to collaborate successfully with the other team members.

Another potential red flag: if you ask a candidate about their life outside of work and they just keep talking about work, they may be a workaholic. Although it might sound nice to have someone like that on your team, people who don't get downtime eventually burn out and are often unhappy. Both will lower their individual and team productivity. Finally, when candidates give you canned answers that you've heard numerous times in previous interviews, they probably lack creativity. Obviously preparing for an interview is essential, but you also want someone who can think on their feet and who answers naturally and with confidence.

Sample Interview Conversation

Let's say that you're hiring a digital marketing associate for your team and are looking for a range of soft and hard skills. To give you a better sense of the personality traits that may appear in an interview and what they might mean, here's a sample dialogue.

> **You:** Tell me about your experience using digital tools to increase leads for your previous company.

> **Candidate:** I've had five years of experience in digital marketing, using tools such as email marketing, social media, and mobile. In one online marketing campaign, we used a variety of social media sites to generate a million impressions, which resulted in half a million dollars in new product sales.

The candidate's response not only succinctly answers your question but also shows that they've been able to use a variety of tools to generate real business results. This is an indication that they're competent and could predict that they'd produce similar results for you.

> **You:** I'd like to hear about a time when you made a mistake at work and what you did to make it right.

> **Candidate:** In my last job, I was unable to meet a project deadline because I was overloaded with work and failed to prioritize accordingly. The first thing I did to

> resolve the problem was to get my teammates together and acknowledge that I'd made a mistake. I then asked them for a few days' extension. Ultimately, I was then able to finish up the project without having too much of a negative impact on my team.

At first you might think that this candidate would be unreliable. But they were honest and explained how they handled the situation with professionalism. Most applicants won't admit their mistakes because they don't want to be perceived as having failed or as not being worthy of the job. But everyone makes mistakes, and honesty is the most important thing in building a trusting relationship with anyone.

You: How do you feel you'd fit into our organization?

Candidate: I plan to use new technologies to advance your organization's marketing efforts, and I want to eventually be a chief marketing officer. I have a passion for working with strong teams, learning from them, and achieving great results through our combined efforts.

This response tells you about the candidate's ambition and motivation and shows that they care about the team instead of focusing solely on their own interests. If you're interviewing someone who has the right skills and expertise, is motivated, and is a team player, why look any further?

You: Do you have any questions for me?

Candidate: What will my day-to-day tasks be if I'm hired for the job?

This is a great question because a job description rarely tells you exactly what you'll be doing on an ongoing basis or how work actually gets done in the organization. This candidate is curious and wants to know what she's getting herself into before accepting an offer. I'm guessing that you can think of a time or two that you wished you had asked a similar question before taking a job.

Creating a Unique Interview Experience

You don't have to restrict yourself to the traditional interview format or setting. When you put candidates in an unusual or unexpected environment, you often get a better sense of their personality and can better gauge their problem-solving skills. You can also see how they adjust their style, posture, and ability to think on their feet. The following are three examples of how to make the interview situation more effective, unique, and useful for both you and the candidate you're interviewing.

The Café Interview

During one of my first interviews right out of school, one hiring manager invited me to go to a café instead of his office. I had never had that type of interview before, but he told me after I got the job that he found that doing interviews in nontraditional places was a great way to get to know candidates and see how they'd perform in a different environment. I still considered it an interview and brought my A game to the café. The next time you do an interview, tell the candidate to come to your office, but then take them out somewhere to see how they adjust and whether the less formal surroundings make them open up and relax.

The Business Challenge Interview

Instead of a traditional interview in which you ask a series of questions, give the candidate a real business challenge to present on right in front of you. The challenge could be something you're thinking about with your team or something you've worked on in the past. This format will give you a sense of how the candidate would perform if hired (and might help your team solve a problem whether you end up hiring the candidate or not). Melanie Chase, vice president of brand marketing at Fitbit, uses this approach when hiring. "I might ask to get a candidate's take on how we bring new user groups into the tracker category or to discuss an approach we are taking in one of our international markets. How the candidate wants to present is up to them and their style. Seeing a candidate present not only gives me a sense of their problem-solving abilities and creativity, but also how they communicate and answer tough questions. The best candidates are those I learn something new from. I love people who challenge our way of thinking, spark an interesting discussion, or show off a unique area of expertise. It's not about getting an answer right, but about demonstrating how they got to an answer."

The Extended Interview

If you're serious about finding the right person for the position you're hiring for, you may need to invest time. Om Marwah, global head of behavioral science at Walmart, is willing to put in hours for a single candidate to test the candidate's stamina and feel them out. He often interviews people for more than four hours—sometimes right through dinner—to assess their capacity to keep going. "I had a candidate that tired me out," he told me. "She just kept going, coming up with fantastic ideas, and had unlimited energy so I hired her on the spot. She was the best person I've hired." Aside from demonstrating stamina, her style aligned perfectly with Om's work style, so it was clearly a great fit for both.

The First Interview

If the first interview doesn't go well, you won't ask the person to come in for a second round of interviews and to meet the team. Good chemistry between an interviewee and an interviewer is a good indicator of whether the work relationship will work.

This means that you should have a pretty good handle on the personality traits and hard skills that you're looking for in a prospective team member. And you should have an equally good (or maybe a better) handle on deal breakers that will prevent you from hiring them or even finishing the interview. Although it's a good idea to have a written list of what you're looking for and what will end things, it's also important to listen to your gut. Quite often we look back and can clearly identify a red flag or two that we willfully disregarded during the hiring process because everything else seemed so great.

Onboarding a New Hire

Onboarding might not sound exciting, but it's extremely important to the long-term success of your new hires. Their first few weeks on the job allow you to set expectations and goals, and that time gives new hires a chance to connect with the key players they'll be interacting with and to learn what you expect from them so they can excel. The more you invest in their careers in the first weeks, the bigger the long-term payoff will be. To be more specific, a successful onboarding program can increase retention by 25 percent and improve performance by more than 10 percent.[20] During employees' first week on the job, they want on-the-job training, a review of your company policies, a tour, and to find a mentor (you or someone else).[21] Let them shadow you so they see what you do every day, and spend time introducing them to your team members and others they'll be working with regularly. (But don't go overboard; too many introductions can be overwhelming.) Instruct them on office protocol and how to reserve a conference room, and give them an employee handbook if your company offers one. Be there to mentor at first so they feel comfortable on your team. Eventually they'll find others to support them as they get situated in your organization.

Onboarding Checklist	
☐	Collect the required new employee registration data.
☐	Provide access to the employee handbook.
☐	Introduce them to the tech your team uses.

(continued)

☐	Orient them with your office facility.
☐	Allow them to shadow you.
☐	Schedule a weekly meeting with them to sync up.
☐	Give them training.
☐	Hold an all-hands team meeting to introduce them.
☐	Communicate your expectations and goals.
☐	Ask them what their career goals are.
☐	Allow them to evaluate your onboarding process.

During the onboarding process, don't overload your new hires with too much information, or they won't retain it and will get stressed out. And don't assume they'll be able to hit the ground running just because they've had prior work experience. When developing an onboarding experience, use technology as a way to increase efficiencies instead of relying on it to immerse the employee in your culture. For instance, offer a mobile-friendly handbook instead of a hard copy, artificial intelligence to answer standard new-hire questions, virtual training options, and a directory of staff so they can familiarize themselves with the organizational structure. While technology can be useful, nothing replaces the human touch, especially when you're onboarding new employees. You should meet them, introduce them to their teammates, get lunch with them, and offer them mentoring support. Interact with them daily to ensure they have everything they need to become fully productive employees over the next few months. Finally, don't fill their calendars with meetings, and urge them to allow themselves plenty of time to learn their new roles and start getting to know their teammates.

Key Takeaways from Hire for Personality

1. **Ask the right questions when hiring.** Seek to uncover the five key personality traits that will help ensure that you'll find someone who fits with your culture and who can perform well long-term. Hard skills are important, but it's personality that can really unite the team and make people collaborate more effectively.
2. **Interview candidates in different settings outside of your office.** You will see how they behave and get to know them more personally. You wouldn't be interviewing them if they weren't qualified and didn't have the right technical skills to do the job. What you're really assessing in an in-person interview is "fit": whether you and your team will get **along with the candidates.**
3. **Spend the necessary time onboarding your new hires.** Help them feel like they belong and like you're fully invested in their careers and want them to succeed. That success depends a lot on how employees are brought onto the team in the first place. Introduce them to their coworkers and to the key players in your organization that they'll be interacting with, and allow them to shadow you so they get a sense of what their jobs will be in the context of yours.

Chapter 8

Engage to Retain

Put the time and hard work and effort into not just forming relationships, but sustaining and continuing and growing the best existing relationships.

—Tom Rath[1]

More than three out of four full-time workers are either actively looking for a job or open to new opportunities. Meanwhile, nearly half of companies can't fill their job vacancies.[2] As a result, we're in a situation that I call "the continuous job search," in which today's workers are only a few clicks away from lining up their next job interview even if they're sitting in an office twenty feet away from you. Because replacing team members is expensive and can really hurt your productivity, the most effective way to retain top people is to support and engage them and give them a positive work experience. Only 4 percent of workers who feel engaged on the job would leave their current employer within a year. But a third of those who feel disengaged are ready to go.[3]

We work in a world marked by tons of collaborative technologies that are designed to boost engagement, productivity, and results. However, over the decade or so that I've been studying workplace trends and employee behavior, I've noticed a definite engagement crisis. And I'm not the only one. According to Gallup, about two-thirds of employees are disengaged at work.[4] Part of the problem is that all that technology—the very things that are supposed to connect us to our teammates—often ends up making us feel even lonelier. Another part of the problem is the rise in remote work, which despite the best of intentions, contributes to a culture of social isolation. And in social isolation, both strong personal relationships and corporate culture suffer.

With nearly a third of companies offering telecommuting,[5] managing freelancers or remote workers has become one of the great challenges for any leader. When we asked employees what they miss out on by not working in an office environment, a third said interactions with coworkers. Anyone who's ever worked remotely or worked with a remote colleague knows that it's hard to stay connected with people when you don't see them and hear their voices on a frequent and regular basis. That's certainly true for me. Even though I'm highly productive working from home, I often feel lonely and isolated from my team. For me and most other people who work remotely, texts, emails, and instant messages on the phone just don't cut it. I'll admit that if I didn't see my business partners face-to-face at least once a year, I'd be a lot less committed to the future of our organization. Without seeing them or hearing their voices, it just wouldn't feel like I was part of the company—even though I'm a partner!

The Virgin Pulse and Future Workplace study uncovered some really interesting data on the effects of remote work on engagement.[6] For example, only 5 percent of people who often or very often work remotely say they can see themselves spending their entire careers with their current employer. Compare that to the almost one-third who rarely or never work remotely and say the same thing.

The Backlash Against Remote Work

As a leader, working from home can be extremely challenging because you can't ever be sure what your employees are doing unless you ask them, and you feel like you're on a team of one even though you're collaborating with others. Your employees may think that you don't care about them or that they aren't part of something bigger than themselves.

A lot of companies have seen how a lack of engagement detracts from their corporate culture and in some cases contributes to poor financial performance. As a result, they're going back to the future. In 2013, Yahoo!, Best Buy, and HP rolled back their remote work policies. More recently, Honeywell, Reddit, and IBM followed suit. These companies had been among the biggest champions of remote work but have since reverted in an effort to rebuild their culture, increase engagement, and get everyone on the same page and focused on the same goals.

"When our company eliminated working from home several months ago, it was disappointing and not fun as a manager to explain to some of my permanently remote employees," says Kiah Erlich, a senior director at Honeywell. "But as a leader who craves human interaction, it has been one of the greatest things we've done. People are actually in the office now. What once was a painful conference call is now a collaborative white boarding session. Instead of more emails, people get out of their chair and walk over to my office. It is a beautiful thing to see, and has not only improved productivity but brought the team closer together."

Why Companies Changed Their Remote Work Policies	
Honeywell	"When employees come into the office to do their jobs, it fosters teamwork and idea sharing. It also helps coworkers make decisions faster and become agile when addressing changes in the global markets."[7]
Yahoo!	"Being a Yahoo isn't just about your day-to-day job; it is about the interactions and experiences that are only possible in our offices."[8]
HP	"We now need to build a stronger culture of engagement and collaboration, and the more employees we get into the office the better company we will be."[9]
Reddit	"While remote work was good for some workers, in the macro scheme of things, the company just wasn't able to collaborate and coordinate efficiently."[10]

I see Kiah's point, but to me, banning remote work is a rather extreme solution—as is a workplace in which *everyone* works remotely. What we need is a mix of both: a place where managers take into consideration each individual's unique preferences, priorities, and needs.

How Engaged Employees Impact Your Team

Remote workers aren't the only ones who may feel isolated or disengaged from their teammates. Over the past decade the workplace has evolved from collections of teams in large commercial office buildings to smaller teams with more discrete projects that can be performed in isolation but are completed collectively by the whole team. As a result, your nonremote employees often work alone at their desks, skip lunch, and have almost as little contact with other humans as they would if they were working at home, where the only sounds they might hear are a car going by or a dog barking. Regardless of your team's size, function, department, or work location, at least one of its members is probably feeling disconnected, disassociated, and lonely and needs to be engaged more.

Employee engagement has an impact on every aspect of your company, from pleasing your customers to strengthening partner relationships to recruiting new teammates. Lack of this engagement can be a massive barrier to your productivity and may slow growth. Highly engaged teams have 41 percent less absenteeism, are 17 percent more productive, and have 24 percent less turnover than disengaged teams.[11] The Hay Group found that companies with a high level of employee engagement generated 2.5 times more revenue than those with less engagement.[12]

As a leader, you want your employees to be emotionally committed to your team and organizational goals. When employees are engaged, they invest more time and effort in their work and relationships. They feel purpose in their jobs and bring the necessary enthusiasm, passion, and energy to their daily activities. Towers Perrin found that 84 percent of highly engaged employees feel as though they can have a positive impact on the quality of their company's products and services, versus only 31 percent of those who are disengaged.[13]

How Presence Impacts Your Team

Being seen is at least as important as being heard. Seeing other people face-to-face is more psychologically engaging and makes you feel more connected. One study of professional office workers examined how passive face time (being seen at work either during normal business hours or off hours) affects the way people are perceived at work.[14] It turns out that passive face time leads observers to perceive employees as either dependable or committed—judgments that observers are completely unaware they're making.

Aside from making leaders seem more dependable and committed, having a presence at the office may actually make you more approachable and empathetic. Mike Maxwell, KitchenAid global category leader of the culinary-global leader program at Whirlpool, is always on the move in the office and believes in the power of being seen. "In the morning, I get up and see my team and check in with others. Touching base with them on work and non-work-related issues. It helps me be more approachable and I get key updates on things I need to know. There is significance in that face-to-face interaction." His team appreciates the effort and care because it gives them time to reflect, be open, and connect.

Your physical absence can affect not only your team's loyalty but also your own career prospects. When you aren't seen, you aren't top of mind when it comes to new projects, promotions, and even bonuses. Those who are there command more attention and are perceived as working harder because their efforts can be observed directly, while remote workers—even if they're rock-star employees—often go unnoticed. As Jack Welch, the former CEO of GE, once said, "Companies rarely promote people into leadership roles who haven't been consistently seen and measured."[15] Being present sends a strong message to others that you're committed to the company and want to lead. It's helpful to your employees to understand the importance of presence and all the great benefits that go along with it.

If you want to build trust with your teammates, you need to successfully engage them—and the best way to do that is to be present. As Amy Cuddy, author of *Presence*, once told me, "Presence allows you to build that trust, because you are saying, 'I'm here, I care about you. I'm listening and what I am telling you to do is not just based on my own personal opinion but what I'm observing and hearing from you.' "[16] You can't have the same level of interaction (or trust) if you're using technology as a crutch to engage with teammates.

The Power of Face-to-Face Engagement in Practice

Facebook's CEO Mark Zuckerberg and COO Sheryl Sandberg form one of the most prominent power teams in our society. Through their combined efforts, they have made decisions that have impacted billions of people. Even though they lead Facebook, which encourages virtual interactions, they rely on face-to-face meetings to be effective leaders and teammates.

Both meet one on one biweekly with no set agenda other than to reflect on new developments at the company. Zuckerberg said these meetings are "a really key way in which we share information and feedback and keep stuff moving forward." From Sandberg's perspective, they're important "because we always know we're going to talk things through and we're going to get on the same page." They could hold these conversations in a Facebook Group or through their messaging service, but these regular human interactions help strengthen their relationship and make them more effective.[17]

Out with the Old Leadership Style and In with the New

In generations past, leaders were often autocratic and spent a lot of time commanding and controlling the behavior of their employees. They believed in rules, practices, and regulations for how work gets done. Autocratic leaders still exist today in every industry, and you may be one yourself. Take, for instance, Lorne Michaels, the creator of NBC's *Saturday Night Live*. Since 1975, Michaels has produced one of the longest-running and most entertaining shows on TV. As an autocratic leader, he still demands the best in every cast member and has final approval over every sketch and scene. Thanks to his hard-hitting leadership style, Michaels helped build some of the great comedic careers, including those of Bill Murray, Eddie Murphy, Amy Poehler, and Tina Fey. Michaels once said, "To me there's no creativity without boundaries," and those boundaries helped spark amazing creativity in his team of writers and on-air personalities.[18]

Steve Jobs had a similar leadership style and built one of the largest companies in the world but was frequently harsh to his employees and had no problem publicly embarrassing them if they said something he didn't like. His style was unwavering, yet his employees were extremely loyal to him because he saw something in them that they didn't see in themselves. While both Jobs and Michaels were quite successful (and Michaels continues to be), I think they'd have a much harder time leading today, when organizations are flatter, more collaborative, more social, and less rules-oriented and allow for greater free flow of information.

Today's transformational leaders are able to adapt to change and new environments, and they prepare their teams to do the same so they can navigate new obstacles with confidence. They create a vision and inspire their teammates to persevere. They encourage the best in others, promote collaboration, and are cheerleaders instead of dictators. As a transformational leader, you must be willing to make the necessary sacrifices and support your team through unexpected challenges. You must be able to communicate your ideas, create trust among your team members, engage with them, and have a strong sense of empathy. And you must create a culture that celebrates success. When we interviewed hundreds of young employees, over 60 percent believed that they lead in that way.[19]

Conflicting Leadership Styles

Older leader: We have a process that every employee needs to follow here, and every project needs to be signed off on by a vice president or above. If you come up with a new idea, don't pursue it without permission. I need to make sure that it fits our guidelines and that your approach is something that has worked previously.

Younger leader: We have a process, but it's flexible, and I encourage you to share new ideas and ways to accomplish this task. Although we have basic guidelines, I think it's important to collaborate and work toward our goals in the most efficient way possible.

Part of creating an engaging culture is empowering and trusting your employees with regard to how they get their work done. If you're always micromanaging them and humiliating them in front of their peers, and you never let them spread their wings, they will feel frustrated and resentful and will soon look for work someplace where they'll be appreciated and nurtured. Sometimes the best leadership style is one that lets go.

In an insightful *New York Times* interview, LinkedIn CEO Jeff Weiner was asked about his definition of leadership. His response echoes transformational leadership. He said, "Simply put, it's the ability to inspire others to achieve shared objectives."[20] He then talked about how vision gives the leader and the team clarity about where to take the business and its products. While Weiner isn't a young employee, he has a lot of experience and insights from being an executive at both LinkedIn and Yahoo!, two prominent technology companies. When I spoke to Drew Houston, the young CEO and cofounder of Dropbox, he said, "If you get results but aren't a team player, it'll be hard for you to succeed here." Houston told me about a transformative leader at his company named Guido van Rossum, the original creator of Python, the popular programming language. Rossum has interns come up to him and tell him why they think Python is inferior to other programming languages, and instead of turning his back on them, he's willing to have a mature debate with them. Rossum is a true transformational leader because he's not stuck in his own ways and welcomes feedback, even if it's not what he wants to hear.[21]

A few years ago I interviewed Biz Stone, one of the three cofounders of Twitter, about his career decisions and influences. I asked him about the leadership of his cofounders and their relationship as they grew Twitter into the behemoth that it is today. "Both Jack and Ev continue to help me grow into a better person in various ways. Jack is extremely supportive of my current startup. Ev gives me more patience than I deserve."[22] Even successful entrepreneurs need a support system and have to leverage one another's strengths. Transformational leaders will power the next generation of companies and make the workplace environment better for everyone.

The Four Ways to Promote Employee Engagement

Through countless research studies and conversations with leaders across a variety of businesses, I've narrowed down employee engagement to four factors: happiness, belonging, purpose, and trust. As a leader, if you can nurture these factors, you'll significantly increase the likelihood that your teammates will stay productive, fulfilled, and committed to your goals. When you hire the right people for the right jobs and empower them with the resources and emotional support they need, they'll be successful. Let's take a look at these factors in greater detail.

Spread More Happiness

Regardless of your industry or company size, it's impossible to ignore the power of a happy employee. Happy employees are more likely to refer new candidates to you, brag about you online, work harder, and remain with you even through the hardest times. My friend Shawn Achor, author of *The Happiness Advantage*, found that happy employees have an average of 31 percent higher productivity, have 37 percent higher sales, and are three times more creative.[23] I know it sounds a little cliché, but when one of your employees is happy, that emotion will spread throughout your team and the rest of your organization.

How to Create Happy Employees

1. Have a serious conversation about work-life balance to show employees that you care about their personal lives and not just their work performance.
2. Do a random act of kindness, like ordering lunch for everyone on a day other than Friday, to show that you care about them.
3. Spend time with your employees, get to know them better, and ask how you can create a better work experience for them.

Create a Sense of Belonging

As humans we have a natural need to be accepted by others, and in the workplace those "others" are our team members. When I was growing up I never fit in, and I had trouble making friends. Without a sense of belonging and with no close friends, I had trouble performing academically, and I wasn't happy. In college I joined a fraternity, which gave me an immediate sense of camaraderie and acceptance because everyone went through the same grueling six weeks of pledging. I didn't have to worry about my social life because it was already established for me after the older fraternity brothers accepted me. That helped me focus a lot more of my time and energy on excelling in my internships and classes, and the results were extremely positive.

In the workplace your employees want to feel that they belong to the team, which is why so many leaders hire for cultural fit in the first place. They want someone who will fit in from day one. When employees have a sense of belonging, their guard comes down and their performance goes up. Belongingness is often overlooked at work because we're so focused on our goals that we don't pay enough attention to how those around us feel.

To create a sense of belonging, you need to make your employees feel they're part of a community every single day—one that helps fulfill their aspirations, supports their well-being, and makes them feel respected. A study by professors at UCLA found that threats to belonging result in an experience that's similar to physical pain.[24] And that's just the beginning. Other studies have found that not belonging or feeling a lack of acceptance can also lead to depression, reduced problem-solving skills, and a decrease in on-the-job effectiveness.

"The interactions that my team has outside of the office are the most important for engagement and retention," says Jill Zakrzewski, customer experience manager at Verizon. "Our bond is created during happy hours, team barbeques, and at the ping pong table, not over a spreadsheet or email. Because we have real relationships outside of the office, we're quick to support each other inside the office."

Tips for Creating More Belongingness

1. Promote belongingness by scheduling social events, having team lunches, and creating an environment in which people feel safe to share information about their personal lives.
2. Hold more meetings that include your whole team so that all feel their voices are being heard and their needs are being met.
3. When you see a teammate who hasn't been acknowledged or included in team activities, make a special effort to reach out and make him feel welcome.

Connect Purpose to the Work

When you have a purpose, you feel that you matter and have a direction to go in. Over the years I've come to realize that my purpose is to help my generation through their entire career trajectory, from student to CEO. My purpose became clear only after years of writing, experience working with companies, and mentoring young professionals. Now, whenever I make a business or life decision I consider whether it works with my purpose. If not, I disregard it. This keeps me focused and keeps me from wasting time on activities that don't help me support my peers in their success. I know it's my purpose because most decisions in my life have guided me to it, and I get excited whenever I think of the possibilities of seeing it through.

I spoke to Michael Porter, famous for the "Porter's Five Forces" framework and a professor at Harvard Business School, about the impact my generation has had on the workforce. He said that today's young workers are "more aware of society's many challenges than previous generations and less willing to accept maximizing shareholder value as a sufficient goal for their work. They are looking for a broader social purpose and want to work somewhere that has such a purpose."[25] Simon Sinek, author of *Start with Why* and perhaps the biggest champion of purposeful work, expanded on Porter's comment. As he explained to me, "Great, inspiring leaders and organizations, regardless of size or industry, all know *why* they do what they do. It is this clear sense of *why* that inspires them and those around them. It is what drives loyalty. And it is what drives their success over and over and over."[26] Purpose has always been important but was often overlooked by previous generations. Now it's one of the hallmarks of a career and one of the key determinants of whether someone will stay with or leave a company.

Tips for Creating Purpose

1. Bring in a customer who has been personally affected by your team's work so that your employees can hear and see the impact of their efforts.
2. Don't just assign work; make sure your teammates know *why* they're doing it and how it will be used to support your organization, your customers, or even the world.
3. Have employees share their accomplishments and the sense of purpose they feel when they come to work every day. This will help identify shared purpose and goals.

Establish and Maintain Trust

Trust is the hallmark of any good relationship, be it work or personal. When your employees trust you, they'll tell you things that you may not *want* to hear but that you truly *need* to hear. They'll be more open with you, provide feedback, and be able to voice issues without stressing about feeling judged or making a fool of themselves. Paul Zak, a researcher at California's Claremont Graduate University, found that countries that have high levels of trust between citizens are more economically successful. In one experiment he monitored his subjects' oxytocin levels. Oxytocin plays a role in social bonding, regulating social interaction. Zak found that when people feel trusted, their brain produces oxytocin, and the more they feel trusted, the more oxytocin they produce. In addition, individuals who felt trusted were in turn more trusting of others. In the workplace, employees at high-trust companies experienced 74 percent less stress, had over 106 percent more energy, were 50 percent more productive, took 13 percent fewer sick days, were 76 percent more engaged, felt 29 percent more satisfied with life, and were 40 percent less likely to burn out.[27] Without trust, employees feel that they can't bring issues to your attention and can't rely on others to help them complete tasks. Ultimately, employees who don't trust others or who aren't trustworthy themselves won't last on your team.

It's hard to establish the kind of trust I'm talking about when you're relying on your cell phone all the time and not having in-person conversations. Texts, instant messages, and emails are technically forms of communication, but they aren't nearly as useful for creating strong relationships. It's a lot easier to trust someone who's right in front of you, speaking with you face-to-face.

Joe Lawrence, director of field training with the US Air Force, says, "Trust is my number-one priority. My formula for trust is to be visible, interested, and involved. I make it a point to get into their classroom for an hour of instruction. I go to their office if I have a question or want their feedback. I make every effort for them to know I value their time as much as I value my own. Then

I reflect on the things I see that I can remove from their path or a resource I can add to make their lives easier."

Tips for Earning Your Employees' Trust

1. Be transparent with your team, letting them know what's really on your mind and keeping corporate speak to a minimum.
2. Admit your mistakes. That makes you seem more human and trustworthy and will make it easier for others to admit their own mistakes.
3. Keep your promises so that people know they can always count on you.

What to Do if You or Someone on Your Team Feels Isolated

As an entrepreneur, one of my biggest fears is feeling isolated and lonely. Whether you are working from home or are the sole person responsible for a project, it's easy to feel really alone and not part of the team or organization even though you are. If you've ever had these feelings in your career, you know that the only solution is to make a concerted effort to reach out to others or ask for help. The key is to redesign tasks and processes so that you (or someone on your team who's feeling isolated) can have more interactions with teammates on a regular basis. Try assigning two people to a project instead of having everyone work solo. This may occasionally delay projects, but it's better for your employees' mental and emotional health. Many Silicon Valley companies strongly encourage their employees to come to, and live near, the office. They've found that more frequent interactions between team members create new relationships and ideas and lead to better problem solving. In our Virgin Pulse study, 60 percent of respondents said that they'd be more likely to stay with their current employer if they had more friends at work. That's especially true for younger workers: Gen Z (74 percent) and millennials (69 percent).[28]

It really all comes down to in-person communication. Of workers who feel that they don't have enough face time with members of their team, 48 percent said they aren't engaged and 78 percent said they often or always feel lonely. In addition, people felt that having more face time with their coworkers would reduce the amount of time they spend going back and forth on email, increase their on-the-job learning, boost their confidence and productivity, improve their relationships with their colleagues, make them more committed to the organization, and give them a better chance of getting a promotion.

How Great Leaders Engage Employees

Engagement isn't something you do once a week or year. It's the interactions you have with your teammates throughout the day, every day. Over the years I have researched the top ways that employees want their leaders to engage with them so they can do their best work.[29] None of them have anything to do with technology (although it could help in certain circumstances). Here are eight practices great leaders apply to engage employees.

1. **Listen to teammates, value their opinions, and create a culture of respect.** Your employees want to share their thoughts with you without having to go through ten layers of management to do so. If they know that you're listening to what they have to say and that you'll actually take action on it (because actions speak much louder than words), they will take you seriously and will continue to share. When employees feel ignored, they'll see sharing their ideas as a waste of time and effort.
2. **Give employees meaningful projects.** They want projects that have an effect on their team, their organization, their customers, and the outside world. No one wants to feel like a mindless cog in an even-more-mindless wheel. They want to see how their tasks add up to larger projects and initiatives and the impact they make. As a leader,

one of your biggest challenges is to tell a compelling story to your employees that will make them fully understand how their daily tasks fit into the bigger picture. Whether your employees would like to admit it or not, a simple biweekly paycheck doesn't cut it when it comes to engagement; people want to feel that they matter. To assign the right projects to the right team members, you need to hire the right people and set expectations so that employees know what they are responsible for and have the tools and support to succeed.

3. **Mentor employees and give them feedback so they can thrive.** Unfortunately, the word *mentor* is overused and misunderstood. For instance, some people say that anyone you learn from is your mentor; others insist that you can have only one mentor in your career. In truth, mentoring is the pairing of two individuals to create mutual support and success. Although many people think that the mentee is the only one who reaps rewards from the relationship, for the arrangement to be a success, *both* need to benefit. Successful mentoring happens when you truly invest in your employees, setting aside time to help those who are struggling, have questions, or could use some guidance.
4. **Model transparency, providing full access to company information so that everyone's on the same page.** Have weekly meetings in which you fill in your team on the latest company news and acknowledge both the victories and the setbacks so they always know exactly what's going on and where they stand. By acknowledging both, your team can find ways to bounce back from setbacks, share in the glory of big wins, and become future-proof.
5. **Coach employees, don't just mentor them.** When you're coaching, provide actionable feedback and advice, using specific examples from your experiences so that

they fully understand what to do. Both the employees and you should keep notes so that you can track progress against mutually established goals and expectations. Michelle Odland, business unit director for Old El Paso & Totino's at General Mills, does monthly or quarterly coaching sessions with her employees because it not only helps their development but helps her do her job better. "To truly learn something is to teach it, so explaining and discussing different development concepts with each employee helps me more deeply appreciate and understand the concepts." To Michelle, a coaching session is a two-way street, and dialogue, which reinforces the value she places on her teammates, gives her critical feedback. It also reaffirms to them that their voices and perspectives matter. This coaching style gets both parties on board and invested in each other, which furthers their development.

How to Coach an Employee

1. Let employees know that you expect them to be accountable for their actions, but show them that you're confident in their abilities to complete the desired tasks. This way, they feel like you trust them to solve problems and that you're interested in their performance.
2. Point out weaknesses or performance issues so that employees can reflect on mistakes they made or skills they may lack. Focus on the problem or behavior that needs to be improved and give them specific examples of times when they could have acted or done the work in a better way. Although we want to improve their issues, don't shy away from recognizing the things they've done well first. People are more likely to take criticism when they are complimented first.
3. Work with employees to solve issues in a collaborative way that makes them feel that they're part of the solution, not just part of

the problem. The idea is to ensure that mistakes don't happen again and to clearly establish that you're committed to helping your employees improve every day, not just this one specific time.

4. Get the employees to agree that the issue exists. Sometimes employees have no idea that there's a problem in the first place and need coaching to help them see the problem.
5. Finally, get the employees' commitment to resolve the problem with mutually agreed-upon solutions. By working together to create solutions, over time you'll be helping them develop the skills they'll need to come up with their own solutions in the future.

---

6. **Always be there for your employees.** Arguments will arise between your team members and other people. Your employees need to know that you have their backs and will support them when they need it most. Let's say one of your teammates is working on a cross-functional team and isn't being treated fairly. If they can't handle the situation on their own, you may need to step in and straighten things out. When you protect your employees, they'll be much more engaged and loyal to you as a result.
7. **Provide opportunities for employees to grow and develop.** Putting yourself in the employees' shoes shouldn't be too hard because you were there just a few years ago. Just as you did earlier in your career, employees want to expand their skill sets and continue to move up in your organization. If you don't find new ways to promote them or challenge them with new responsibilities, they'll become bored and disengage. So, sit with them and discuss their individual career plans so you can tailor opportunities and challenges to their wants, desires, strengths, and abilities. If you're not able to find new,

appropriate challenges for them on your team, you may need to help them find a new role with another team; if you don't, they'll leave anyway.

8. **Celebrate teammates' achievements, making them feel good about their jobs and being part of something special.** This means that if someone does a bang-up job on a presentation, gets promoted, or has a birthday or a baby, make a point to say something to that employee and the group about it. It's a great way to show that this person matters and that you care on a human level. When you do this, you're contributing to a culture of joy and recognition, and that teammate will acknowledge others when they do something great or have a special moment. Although you could probably get away with a text or an email, be human and have an actual conversation. The genuine emotion you'll give off is much more powerful than anything digital.

How to Engage Remote Workers

With more and more employees choosing flexible work arrangements, chances are good that you're managing one or more remote workers now or you will be in the not-too-distant future. But although some people might think that managing someone they rarely see would be easy, it actually takes a lot of skill. Collaborative technologies can be useful for maintaining contact with your remote workers, but it's your responsibility to create a more personal relationship with them by phone, in-person visits, or videoconferencing.

When remote teams are disengaged, it's almost always the manager's fault. In one study, a fourth of virtual teams were ineffective due to poor management. Highly effective teams, however, had management who promoted accountability, motivation, purpose, process, and of course, relationships.[30] Managers of the most successful teams have a consistently open line of communication with their employees and ensure that those employees know what's happening in the organization and have the tools they need to succeed.

Remote workers face a number of unique challenges. First, they're much more likely to feel isolated, lonely, and depressed because they have less human contact than your other employees. Second, they have many more potential distractions around them, such as noisy people in a coffee shop or their kids playing video games in another part of the home. Third, although their intention behind working remotely was to have more flexibility, they often end up losing track of time and working around the clock. When you're in an office you have a more defined workday because you see people coming and going at around the same time every day.

You won't be able to overcome all of these challenges, but you can definitely help by giving remote workers regular attention. Be sure that your goals and your telecommuting policy are clear so there won't be any issues later. Explain how important remote workers are to the whole team's success and ask for regular updates so remote workers get in the habit of checking in with you. Make sure you aren't just relying on high-tech collaboration tools to engage with them. Instead, schedule at least one weekly phone meeting or videoconference so you can maintain an emotional connection with them and ask whether they have any project updates or need help with anything. Issues that don't get resolved quickly can fester and quickly become much bigger problems.

"Whenever there's a lag of communication across a remote workforce, people fill that void with assumptions about what's happening or not happening in the business, and with rumors, and they feel very disconnected and very uneasy about the business," says Rajiv Kumar, president and chief medical officer at Virgin Pulse. "Constant communication—and never leaving remote employees feeling like they're on an island—is critical," he adds, "whether it be company-wide town halls, webinars, emails, PowerPoint presentations, updates on health of the business, social emails, [or] announcing accomplishments."

If you want to take things a step further, consider mandating that your remote employees come to the office a few times every year. While this may seem harsh, it's for everyone's benefit because the extra face time will result in deeper relationships. Antonio McBroom, the chief euphoria officer at Ben & Jerry's, told me that he requires that remote workers "travel to their primary place of business quarterly to reinforce our team dynamic and the importance of the work we're doing." In addition, "any professional development or beyond-the-job opportunities that we have at our home office, we also try to incorporate the parallel for remote workers."

Empower your remote workers by giving them plenty of autonomy and control over the projects you assign them. Micromanaging may seem like a good way to engage them on an ongoing basis, but many people telecommute for the freedom and opportunity to do projects independently. As long as they're producing quality work, give them the flexibility to do things their way. I know it sounds a little counterintuitive, but this will make them feel more included and will encourage the rest of the team to engage more with their remote colleagues.

As you know, my goal in this book is to push you to increase the degree of human connection you have with your teammates. However, it would be absurd to ignore the powerful role technology can play in facilitating communication and collaboration. When you have remote workers, technology can be a vital asset—as long as you use it properly. If you select the right tools, train your employees on how to use them, and then leverage those tools when collaborating, you'll be more successful. If you champion a tool and then fail to use it, your team will eventually stop using it as well. Some tools can limit the often-overwhelming flow of emails we receive every day and enable us to communicate more quickly and efficiently with people who aren't at our location. Use these tools for real-time dialogue and to promote your weekly staff meetings for both your traditional and your remote employees.

Finally, try not to treat remote workers differently than your traditional employees. A lot of leaders pick favorites (usually people they see every day), leaving remote workers feeling they can't succeed or aren't getting as much attention. That's why it's so important to show empathy and concern for your remote workers' well-being and to ensure that they're achieving their personal and professional goals, not just the team goals you've set for them. Nawal Fakhoury, learning and employee experience manager at LinkedIn, takes this message seriously. "Managing remote teams is similar to managing regular teams but will require greater emphasis on building trust, fostering communication, and implementing team processes," she told me. Her solution? "I spend double the amount of time with my remote employees as I do with my teammates next to me. I hold regular 1:1 meetings where we focus on what's going on in their life outside of the office, learning about their work environment, and always asking, 'What can I do to support you?' All meetings are conducted via videoconference so we can make eye contact and pick up on the body language to truly get a sense for how the remote employee is doing since I don't have the benefit of seeing them in their element."

Remote working can be a benefit, a choice, or something that's forced on someone by circumstances beyond his control. So, take some time to understand exactly why your remote employees are in that position, and do everything you can to create a positive employee experience for them. At the same time, it's important to understand your remote workers' communication preferences. Vivek Raval, head of performance management at Facebook, always asks of his remote workforce, "Do they like to video chat? Are they the kind of person who appreciates informal phone conversations not tied to specific meetings or tasks? What can we do to make them feel like part of the team? Can I schedule a visit to where they are located to show a commitment and share some important face-to-face time?" By asking these questions—and more important, listening to the answers—you'll be able to pick the right communication channels for each person's preferences.

Key Takeaways from Engage to Retain

1. **Become a support system for your teammates.** Instead of trying to push policies on your teammates, empower them to take on new challenges. Encourage them to become the best version of themselves and support their own ambitions.
2. **Coach your teammates on how to tackle problems.** Use your own experience as a guide to work with them to codevelop solutions so they'll feel that their voices are heard and that you trust them.
3. **Pay attention to your remote workers.** Treat them the same as your traditional ones so that they feel they're valued and are instrumental to the entire team's success.

Chapter 9

Lead with Empathy

Be kind to the people around you. Teammates/coworkers are family, you spend a lot of your life with them, so treat them with respect and make sure you're creating a positive vibe in the workplace.

—David Ortiz[1]

We live in a chaotic, stressful, and unpredictable world where we're bombarded every day by stories of murder, greed, bullying, terrorism, sexual harassment, poor labor practices, and ethics violations. Reading or watching these negative and disturbing stories typically brings up feelings of sympathy for the victims. If you've been a victim yourself, however, you may experience a completely different—and much more powerful—feeling: *empathy*.

When we hear the word *empathy*, many of us simply nod without fully understanding what it means, or we confuse it with sympathy. The two words are sometimes used interchangeably, but they're quite different. Sympathy is a feeling of sadness or pity for the victim, while empathy is the ability to understand someone's feelings as if they were our own—in many cases because we've actually had them.

Empathy is the most important ingredient in successful, long-term relationships with your teammates, family, and friends.

How Technology and the Media Are Killing Empathy

Every generation has access to more technology and information than their parents did. A child born today will probably know what virtual reality is by age two, whereas I just learned about it a few years ago. The cell phone I had back in college could barely take pictures at all; today, you can shoot high-definition video and upload it to the cloud.

While we may think that technology advancements are cool, they impact our ability to be empathetic and impede our ability to develop the deep relationships we need to thrive as leaders. And the process starts during (or before) adolescence. Dr. Gary Small, a professor of psychiatry and aging at UCLA, said, "The digital world has rewired teen brains and made them less able to recognize and share feelings of happiness, sadness, or anger."[2]

MIT professor Sherry Turkle explains that technology prevents us from learning how to be empathetic. "It's not some silly causal effect, that if you text you have less empathy, it's that you're not getting practice in the stuff that gives you empathy."[3] Turkle says that when you apologize face-to-face, you see the other person's body language and potential tears and you know how upset she is. And the other person gets to see, from your body language and facial expressions, that you have genuine compassion and are truly sorry. Texting the words "I'm sorry" creates no emotional connection at all. In fact, it may end up making things worse. A girl I was dating broke up with me by text an hour before I was to make a big keynote speech to thousands of people, leaving me feeling empty and confused. She could have gotten together with me in person or even called, but to her it was easier to text. Easier isn't always better—at least not for everyone involved. It's hard to develop deeper relationships with others when you're behaving in a way that actively interferes with those relationships.

Although I didn't have a smartphone when I was a teenager, I was a video game nerd. However, I was good only at fighting and strategy games. As much as I hate to admit it, playing those video games probably changed the way I see the world even today, for better and worse. I think I excel at solving problems, but I may be a bit less empathetic because I got used to seeing so much violence—at least in the games I played. And I'm far from alone.

I'm not saying that we should never focus on our phones and other devices. There's no question that they're essential communication tools. But when technology interferes with our ability to maintain eye contact with others, notice nonverbal cues, and socialize with our fellow humans, we've got a real problem.

The media contributes, too. When every day we see people all over the world dying, we develop what's sometimes called "compassion fatigue" and get used to a "new normal" that really should horrify us. It's difficult to be empathetic when there's a new tragedy every minute. How do we even choose whom to empathize with today when we know that something worse will happen tomorrow? We get so used to bad things happening that we forget to feel for others. And not being able to feel for others will make it hard for you to be an effective leader.

Reducing the number of human interactions we have has also reduced our ability to empathize with others, largely because we're losing our ability to experience the kinds of emotions that have traditionally bound us together on the most personal level. Every side conversation, meeting, or work party is an opportunity to show emotion, be vulnerable, and be compassionate to others. Those interactions—and the opportunities to express those emotions—can have an astronomical effect on our work lives. Keep in mind, empathy isn't about feeling bad for someone who's going through a major crisis; it's the daily expressions and gestures that keep people feeling secure in their jobs. Danny Gaynor, who works on the Narrative, Innovation, and Executives team at Nike, says, "Technology is supposed to make us more empathetic by expanding our ability to digest more information. But in the workplace, technology often diminishes empathy. All too often, people deliver tough news from behind the emotional fortress of a harshly written email—instead of having the empathy to offer feedback face-to-face. All too often, people who sit together will deliberate via chat or remote link—instead of meeting together, allowing people to convey their passion with the full range of humanity. We need to recognize that there are human cues—body language, eye contact, and tone of voice—that are fundamental to communication. Technology has not yet provided a substitute for that. So, when possible, treat people like people."

To be fair, technology isn't all bad. Ilona Jurkiewicz, vice president of talent and development in early careers at Thomson Reuters, gave me a number of examples. "Every day people connect in a very deep and powerful way over dating apps. In some instances, people talk on apps like Tinder for weeks at a time and cultivate deep and meaningful relationships purely from text," she says. "When I was in my teens I had a very deep personal connection with someone purely based on communication through phone calls, online messaging, and written letters. We didn't meet for many years, but in fact through these communications (not face-to-face!) we forged an incredibly deep bond that has carried over nearly 18 years.... So, if we're able to forge relationships that matter to us through texting on dating apps, snail mail pen pals, or text messages.... Why can't we do that at work over video conferencing or phone calls? I definitely think we can, and I see it happen all the time. It just takes effort and work." Fair enough, Ilona. But as we'll see later in this chapter, even she believes in the importance of in-person communication for building empathy.

I've been a victim of bullying many times in my life, from getting shoved in a locker in middle school to being cyberbullied after publishing my opinions on my personal blog a decade ago. Constantly being made fun of by my own peers back then continues to affect whom I trust and how I relate to others today. When we've endured failure, injury, family death, harassment, bullying, or something else, we're better able to relate to others' setbacks and crises. When tragedy strikes other people, your ability to put yourself in their shoes, to truly understand what they're feeling, may not immediately resolve things, but it'll bring you closer together.

As Maya Angelou famously said, "People will forget what you say, people will forget what you did, but people will never forget how you made them feel."[4]

Unfortunately the relentless barrage of news stories about tragedy and suffering and our technology addiction, which too often disconnects us from our fellow humans, have combined to reduce our capacity to empathize with others. In his now-famous commencement speech to Northwestern University graduates in 2006, then senator Barack Obama said, "We live in a culture that discourages empathy. A culture that all too often tells us that our principal goal in life is to be rich, thin, young, famous, safe, and entertained. A culture where those in power all too often encourage our most selfish impulses."[5] We are so distracted by social media updates, sports, and movies that we have lost track of what's truly important to our long-term happiness and success: fruitful relationships with our family, friends, and those we work with. Fancy cars, gold watches, mansions, and other material things might motivate you, but they'll also steer you away from the relationships that fulfill you.

Our empathy compass is so broken that *Sesame Street*, a show that many children grew up with, had to hire actor Mark Ruffalo to act out several scenarios to explain it to children. "Empathy is when you are able to understand and care about how someone else is feeling," he said. The lack of empathy that is epidemic in our culture hurts our relationships, both professional and personal.

When Leaders Lack Empathy, Employee Performance Declines

In a culture that promotes individualism (as ours does), it's tempting to see empathy as a weakness, a defect, an inferiority, or a manifestation of incompetence. It's none of those things. And leaders who see empathy as anything other than a positive trait do so at their own peril.

Unfortunately, everywhere we look we're witnessing the polar opposite of empathetic leaders. Instead of genuinely caring for and relating to the people they lead, too many leaders are apathetic, narcissistic, self-serving, power-hungry, and completely misguided. Instead of creating a loving and enriching culture, they wreak havoc on their teams (and our society as a whole), often without enough self-awareness to see (or care about) the harm they've caused.

It's no wonder that today's workers are restless, stressed, and struggling. While many are having trouble paying for housing and groceries, CEOs are making, on average, 271 times more than the typical person who works for them—and that gap is growing.[6] Sexual harassment in the workplace is a growing issue, and although the stereotypical case is a man harassing a female subordinate, it's really a question of power; those with more of it—whether they're male or female—prey on those with less, regardless of sex. After Uber's HR team ignored her reports of sexual harassment, Susan Fowler came forward with her allegations, and twenty employees ended up being fired.[7] Those employees undoubtedly acted badly and need to be held accountable, but the larger problem is the culture of sexism that was created by the CEO. Harassment doesn't just happen at the office, either. After interviewing more than four thousand adults, Pew Research found that 41 percent of people have been personally subjected to harassing behavior online, and two-thirds have witnessed the behavior directed at others.[8]

Workplace bullying is another huge issue that we don't hear nearly enough about. According to the Workplace Bullying Institute, more than sixty million US workers are affected by on-the-job bullying. That's horrifying enough, but what's worse is employers' responses: 25 percent do nothing, and 46 percent do "sham" investigations. Only 23 percent help the victim, and just 6 percent punish the perpetrator.[9] Leaders who bully or harass or who allow it to happen unpunished on their watch cost their companies billions in diminished worker creativity, higher turnover, lower morale, increased absenteeism, lost productivity, increased workers' compensation premiums, poor physical and mental health for workers, workplace accidents, negative publicity for the firm, and of course, lawsuits and financial settlements. All of this is brought to you by leaders who lack empathy.

And let's not forget about greed, a trait often exhibited by those who lack empathy and ethics. A great example of this is Wells Fargo Bank, one of the largest financial institutions in the world, where from 2002 to 2017, financial advisers created millions of fake accounts in their customers' names. As you can imagine, when customers started getting billed for fees on accounts that they'd never opened, they weren't terribly happy. Eventually many of those advisers and their supervisors were fired, the board pushed out the CEO, and the company paid out $142 million in penalties. But over that period the CEO made millions and the bank made billions.[10] Clearly there was a financial incentive for Wells Fargo to go to the dark side; the money it made far outweighed the fines and penalties it had to pay. That's essentially the same logic Ford used in the 1970s when figuring out how to deal with its Pinto automobiles, many of which caught fire in relatively minor accidents and injured or killed dozens. Company executives made a coldhearted decision that it would cost less to settle lawsuits than to recall more than a million cars and pay to repair them.

Unfeeling leaders may also be less likely to care about the safety of their employees. That 271-times salary gap can make it difficult for a CEO to relate to their workers or to see them as anything other than drones that exist simply to make the company money. I'm sure you've read plenty of stories about workplaces where conditions are unsafe or unsanitary and employees have been injured or killed. Empathetic leaders are concerned with their employees' physical health.

Fortunately not everyone can relate to dealing with leaders who have low (or no) ethical standards or to being bullied. But everyone has had to deal with corporate politics at some point. Maybe your manager took credit for your work, or you were passed over for a promotion by a low-performing teammate who had a better personal relationship with the boss. A lot of leaders feel that they need to be involved in office politics to be successful.

And 60 percent of workers in a study by Robert Half agreed, believing that involvement in office politics is at least somewhat necessary to get ahead.[11]

Of course not all office politics is destructive—and no office could run without at least some politics. But I advise you to stay far away from the politics of gossip, favoritism, and scheming to get ahead by pushing down your teammates. Those strategies may seem like good ideas in the moment, but they'll come back to bite you. In addition, given how often we change jobs, you never know whom you'll be working with or for in your next job or the one after that. Burning bridges is never smart.

Narcissists R Us: We're Too Focused on Ourselves and Not Enough on Others

Sara H. Konrath, a professor at the University of Michigan, has chronicled a steady decline in college students' self-reported empathy scores since 1980. At the same time, narcissism scores have never been higher, according to Jean M. Twenge, a psychologist at San Diego State University.[12] She analyzed data from fifteen thousand college students and found that there's a relationship between birth year and narcissism score, with people born more recently expressing more narcissism than their elders. Twenge also found that despite claiming that we value community service, most of us would rather watch TV, play games, or do something to please ourselves than help others.

Social media contributes to the problem, too. University of Würzburg professor Markus Appel analyzed fifty-seven studies comprising more than twenty-five thousand participants and found that there's a link between the number of friends you have on social networks, the number of photos you upload, and narcissism. The more active you are on social media, the more self-obsessed you become and the less you care for others. We're so obsessed with status updates, likes, comments, and shares and so focused on being noticed and getting approval from others that we've lost our ability to be empathetic.

Helicopter parents (who hover over their children, protecting them from any and all inconvenience), snowplow parents (who remove all obstacles from their children's path), and other similar parents have contributed to the narcissism epidemic by never allowing their children to fail. They're more focused on controlling their children and staying involved in their lives and have less regard for their emotional state, their feelings, and their long-term behavior. If you get treated like you're the center of the universe long enough, you'll start to believe it, which is what's happened to far too many young professionals. And if that isn't the definition of narcissist, I don't know what is.

On the other hand, parents can protect their children against narcissism. Mine, for example, are the most caring and giving people I know. They helped me through many hardships and let me fail and learn from my mistakes. But I always knew they were there for me. I believe the way they raised me is a big part of why I spend so much time helping others.

Assessment: Measure Your Degree of Empathy

Now that you've seen how important empathy is and that many people struggle with it, let's take a look at how empathetic you are. Self-awareness is the first step to being more open, relating to others, and getting in tune with your emotions. Answer the following questions "yes" or "no." If you have more than three "no"s, you probably need to learn to express more empathy as a leader.

1. When I hurt someone's feelings, I apologize and admit my mistakes.
2. When the people around me are upset, I become upset.
3. When I see someone else being treated poorly, I get angry and want to help.
4. When there's a disagreement, I try to understand everyone's perspective.
5. When someone on my team does great work, I give that person credit.

(continued)

6. When someone is criticized, I try to imagine how I would feel if I were that person.
7. When people are happy about their success, I feel happy for them.
8. When I see those less fortunate, I have compassion for them.
9. When I'm working with my team, I genuinely care about them as more than just employees.
10. When someone cries, I instinctively want to help.

Empathy in Action

The lack of empathy we're experiencing in our society has created a renaissance of new teams, groups, and companies that are trying to bring compassion back. Since 2015, one of my friends and fellow leaders, Chris Schembra, has been hosting dinners for his friends and colleagues. During the dinners Chris is in the kitchen making his special pasta sauce, while calling in each guest to help him with a specific activity, like slicing, cooking, or serving the food. Then, at dinner, they go around and introduce themselves, not just by saying their names and companies, but by telling the others about people who have had a positive impact on their lives. At almost every dinner, at least one person will actually cry because they are in an environment that feels so safe they can fully express the emotional impact of their story. I talked about my father and how despite not respecting him when I was a child, I've come to understand how supportive he has been of me both personally and professionally. "When people get into the habit of listening to the feelings and perspectives of those around them, they start getting the understanding that these people can teach them everything they need in life," says Chris. This type of understanding, connection, intimacy, and empathy wouldn't be possible in a less human, more tech-device-based format.

While Chris has created an experience to promote empathy at his dinners, Nandi Shareef, who is part of the people development team at Uber, had to lead with empathy in a difficult situation. A peer on her team was struggling with feeling worthy and had begun to question her ability to contribute to her position. Nandi's way of empathizing with her was to take her for drinks and spend time understanding why she felt that way. Nandi then affirmed her teammate's contribution through facts and anecdotes about what she'd given to the team. "Then, I asked her to come home and reflect on things that make her happy, bring her joy, and make her feel worthy, and to use those as regular practice whether things are going right or are going to hell in a handbasket. Ever since, she's walked around the office with her head held high, and she has the results to prove that this shift worked for her."

While Nandi was clearly leading with empathy, sometimes leaders themselves need empathetic teammates to get through tough situations. Jessica Latimer, director of social media at Alex and Ani, experienced this firsthand. While she was going through a divorce, her father was very sick. Jessica wanted to keep her work and personal life separate, so she minimized her sharing of those personal details with her team. After months she finally told two team members what had happened in her life. "One of my team members was so sincere in her response, and the next day she came in with a card and necklace as a gesture that I was going to get through this," she said. "That moment made me realize that it is important to maintain boundaries, but also that we are all human and sometimes you need to let people in." When it comes to empathy, small gestures and open conversations will make your relationships stronger.

Even the most prominent entrepreneurs in our society—despite their fame and fortune—understand the importance of empathy. When not being the inspiration for Robert Downey Jr.'s Iron Man, Elon Musk personifies empathy. Over the years Tesla's working environment has been less safe than the industry average. Musk says that safety is the top priority at his company and showed it by reducing excess overtime (which was linked to high injury rates) and writing a heartfelt letter to his employees. In his letter, Musk not only acknowledged that the safety problem existed and offered to meet with every injured person, but also promised to step onto the production line and do the same tasks his employees were doing. Best of all, he insisted that his managers do the same. What a wonderful way to demonstrate empathy and lead by example!

Elon Musk's Empathetic Letter to His Employees

No words can express how much I care about your safety and well-being. It breaks my heart when someone is injured building cars and trying their best to make Tesla successful. Going forward, I've asked that every injury be reported directly to me, without exception. I'm meeting with the safety team every week and would like to meet every injured person as soon as they are well, so that I can understand from them exactly what we need to do to make it better. I will then go down to the production line and perform the same task that they perform. This is what all managers at Tesla should do as a matter of course. At Tesla, we lead from the front line, not from some safe and comfortable ivory tower. Managers must always put their team's safety above their own.[13]

Here's another example of empathetic leadership in action: Madalyn Parker, an engineer at a small tech firm, wanted to take a mental health day and sent an email to her team saying, "Hey team, I'm taking today and tomorrow to focus on my mental health. Hopefully I'll be back next week refreshed and back to 100 percent."[14] Madalyn's manager responded in the most positive way: "Hey Madalyn, I just wanted to personally thank you for sending emails like this. Every time you do, I use it as a reminder of the importance of using sick days for mental health—I can't believe this is not standard practice at all organizations. You are an example to us all, and help cut through the stigma so we can all bring our whole selves to work." The manager not only understood what his employee was going through, but also commended her for her honesty.

By having empathy, leaders can account for the individual needs of their teammates, and this can make them feel more secure at work as a result. Jason Gong, diversity programs specialist at Pinterest, appreciates his managers because they take into account his individual needs, style, and overall health. "I have a manager that really supports working from home and taking time off for mental health days when necessary. My work can be very taxing emotionally and self-care is key for sustainable success and impact in my line of work." If an employee needs to get into work a bit later, telecommute one day, or care for a sick parent for a week, we need to be understanding of that and make accommodations any way we can. It's our role as leaders not just to help the team succeed at a high level, but also to care for each individual's needs.

A Tale of Two Very Different Leaders

Both Steve Ballmer and Satya Nadella have been the CEO of Microsoft, leading a global organization of more than 100,000 employees. While they've shared a job title, their leadership styles are completely different. Ballmer would go into the office and in the most direct way possible, tell his team about all the things they were doing wrong.

Nadella is the current CEO and leads with a more empathetic approach, believing that humans are wired to have empathy and that they desire harmony in their work. While Ballmer was demanding, Nadella wants to understand where employees are coming from so that he can create a better environment for them. Nadella learned some powerful lessons about empathy when his first child was born with severe cerebral palsy. His wife gave up her career to care for their child, and he realized that to be a better father and husband, he needed to put himself in his child's emotional shoes.[15] This personal experience has brought more humanity into the offices of Microsoft and the products it creates.

Empathy Translates into Real Business Results

At this point some of you may still think that the whole concept of empathy sounds a little too touchy-feely. Well, think again. Even the toughest people on the planet—the Navy SEALs—learn the value of empathy in building teams. To be successful in combat, you need to have a strong support system that's built on trust, which can't exist without empathy. Kwong Weng, who joined the SEALs at age twenty-three, said that he was able to endure all the hardships because he had an emotional connection with his team. Just knowing that he had others who would help if he encountered an obstacle was what kept him going. "When the water was too cold, my buddy would encourage me to go in. When things would get tough, he would say, 'This will pass, just continue,'" Weng said.[16] While the challenges that you'll face inside your company are most likely significantly less life-threatening than what the SEALs face on a daily basis, the point is that shared empathy can get people through anything.

Empathy is at the core of who we are, and it has major business significance. The Consortium for Research on Emotional Intelligence found a correlation between empathy and increased sales. Empathy can boost productivity.[17] A study of radiologists found that they give more accurate, detailed reports when case files include a photo of the patient.[18] And fund-raisers bring in more donations when they tell prospective funders about how the money they give helps scholarship students.[19] According to the Management Research Group, leaders who score highest on empathy are seen as more ethical and more effective.[20] Unfortunately there's a shortage of empathetic leaders. In a report called the "Workplace Empathy Monitor," Businessolver found that only 24 percent of Americans believe that organizations are empathetic, 31 percent of employees believe that profit is all that matters to the organization and that their employer doesn't care about them, and a third of employees say they'd change jobs for equal pay if their new employer were more empathetic than their current one.[21]

Leading with empathy helps you . . .

1. make better strategic choices because you understand where your teammates and employees are coming from,
2. resolve conflicts with care and compassion because you know—and can read—people better,
3. convince your teammates of your point of view because you understand theirs,
4. predict others' actions and reactions because you know what they've been through, and
5. motivate others because you'll have taken the time to understand the things they care most about.

Becoming an Empathetic Leader

To train yourself in how to lead with empathy, take small steps. Set aside time to speak to one of your teammates, and at the beginning of the conversation, just ask how they're feeling. This is an easy, low-stress, direct way of starting an emotional conversation. There's a big difference between asking, "How are you doing?" and "How are you feeling?" The word *feeling* elicits emotion, whereas *doing* is more activity based. Your goal is to get closer to being open instead of using technology to check in.

If you aren't comfortable with the word *feeling*, that's okay. The point is to ask questions that elicit honest answers—questions that can't be answered, "Fine." Sam Worobec, director of training at Chipotle Mexican Grill, used to ask questions like, "How is this project going?" and "Do you have the resources you need?" But over time he switched to deeper questions, such as, "Is the workload too much?" and "I know you have a lot going on at home. Are you handling it okay?" Early on, he was afraid that he was getting too personal, but the results were amazing. "I now have a team that openly talks about what's going on at home while at work. Not the gory details of our lives, but enough that everyone knows what we're all dealing with," he says. "It makes it so much easier to say,

'I'm going through a hard time at home right now,' or, 'We're in the middle of buying a house, so I'll be out of the office a bit more.' Before, our home lives were secrets that we kept from each other and raised a lot of questions when people would struggle or go missing due to things happening in their everyday lives. Now we can empathize, commiserate, and celebrate much more authentically than before."

As great as empathy is, it's still important to have boundaries between work and personal life. "I don't think it is generally useful to have open-ended empathetic discussions at work, as they are often grounded in, 'my job is too hard' or, 'I have too much work to do,' or 'I am too busy,'" says Stephanie Bixler, vice president of technology at Scholastic. "While I feel these feelings myself, I don't think this is a productive way to approach work, as self-pity never gets anyone anywhere. However, for personal matters that are impacting individuals on my team, I think it is highly impactful to have those emotional/empathetic discussions if the individual is interested in doing so." If the individual is *not* interested in engaging in empathetic conversations, whether they're about sexual harassment or death in the family, don't push the issue. But if employees decide to open up to you, listen wholeheartedly.

Bottom line? Know your employees and take your cues from them. "Managing a team of software engineers, the frequency of conversations that require a box of tissues is pretty low," says Sam Violette, manager of mobile and emerging technologies at Land O'Lakes, Inc. "That said, it's so helpful to have conversations about what's happening in people's lives outside the workplace. If I know someone's father is in [the] hospital or that they're selling their home and swamped with everything that comes with that process, I'll know I need to ratchet down that employee's workload. Your best employees will put their work ahead of almost everything else, and sometimes it is your job as a manager to take that decision out of their hands."

During your discussion, you need to do three things:

1. Show that you care by putting your phone away and turning off your alerts. This might sound trivial, but Virginia Tech researcher Shalini Misra found that simply putting a cell phone on the table or holding one in the hand reduces the feelings of "interconnectedness" and empathy in couples.[22] The instant you look at your phone, you're putting up a barrier that will undermine your relationship with the human in front of you and will pretty much guarantee that he or she will be less likely to want to speak with you about personal things in the future.
2. Listen without interrupting.
3. Demonstrate that you understand by summarizing what you think you heard. But don't just parrot back your employee's words. You're probably already familiar with UCLA professor Albert Mehrabian's finding that only about 7 percent of what we communicate is contained in what we say. The other 93 percent comes from our tone of voice and body language.[23] So, pay close attention to those things. If you just listen to words, there's a good chance that you'll miss the essence of what your employee is trying to communicate to you.

When leaders in our organizations display humility and weakness, they become more relatable. In a study by the Center for Creative Leadership, the authors found that transformational leaders need empathy to care about the wants and needs of their followers. Empathy is also positively correlated to job performance. The more you show your compassion and your willingness to help your teammates when they're struggling, the harder they'll work for you, and the more committed they'll be.[24]

We all want to feel that we are important and that we matter. Knowing this, leaders should treat people like they are important and give everyone a fair chance to demonstrate their abilities and showcase who they actually are. Instead of stereotyping employees, include them and make them feel they are part of your team community. In many cases all it takes to do that is a face-to-face conversation. Amit Trivedi, CP infrastructure and analysis manager at Xerox, told me about an experience he had. "One of my team members was skeptical of senior management strategies. In order to understand the basis for their skepticism, I met with the individual for a one-on-one discussion," he said. "During this interaction I learned about how their past work was not appreciated, their feedback to the management was ignored, and they were left feeling like they didn't add much value to the group. I was able to assure the individual that I am as much committed to recognizing the effort put by an individual and taking feedback into consideration as I am committed to delivering our projects. This conversation would not have had the same experience and outcome had it been done via email or even a phone call."

Last but certainly not least, offer help and guide your teammates without asking for anything in return. That's an act of empathy because you're demonstrating that you're willing to invest in others, not just yourself. You're also generating some positive long-term karma (if you believe in that kind of thing). When you do something selfless for others, their natural reaction is to want to do something for you in return.

Becoming a more empathetic leader isn't going to come easy to everyone. (If it did, a lot more people would do it.) But if you work at it, you'll get there. That's exactly what happened with Ilona Jurkiewicz, vice president of talent and development in early careers at Thomson Reuters. "It was hard for me to build meaningful connections with team members when I first started managing because it felt superficial and forced. It wasn't until I learned the art of empathetic questions that I grew to love it," she told me. "When I was trying to learn the skill, I made a system out of it. I always carry notebooks at work, and use them to write down notes. On the inside front cover, I decided to write 5–6 questions that I could always reference at the start of a 1:1. These were personal questions that helped me understand another person's point of view, and to open up a deeper dialogue. So, every time I had a 1:1, I would make sure that I asked two or three of these questions. I realized that by forcing myself to do it I would eventually become more comfortable building relationships at a distance and it would become a learned habit. And, guess what? A year or two later, it is now natural, enjoyable and I relish the opportunity to find moments to empathize. I also think it has made me a much better leader."

Where You Can Show Empathy at Work, and How You Can Do It

Situation	How to Handle It
Your employee has had a death in the family.	Tell them you're sorry about their loss and that you know how it feels. Then give them as much time off as they need to recover.
Your employee is struggling to complete a project.	Ask what they're struggling with and how you can help. You could support them by working with them, training them, providing additional resources, or even reassigning the project. Remind them that we all struggle and that there's nothing wrong with asking for help.
Two employees are arguing with each other.	Meet individually with each employee and listen carefully to both sides of the story. Once you understand what the conflict is about, set up a three-way meeting and try to get the employees to see things from each other's point of view. There's a good chance that that will be enough to get them to resolve their problem on their own.
Your employee is overly stressed out.	Let them know that stress is normal and suggest that they take time to hit the gym, go for a walk, or take the morning off. This won't be the first or last time you'll have to deal with a stressed-out employee, so set a clear precedent that lets your whole team know that you value their mental health and that you'll support whatever they need to do to feel more relaxed.

Leading with Empathy in the #MeToo Era

Sexual harassment has always existed in the workplace, but in 2017 it became a headline issue after a series of sexual misconduct allegations took down a number of high-profile people. After the allegations against former Hollywood mogul Harvey Weinstein, Tarana Burke created the #MeToo hashtag, which was popularized by Alyssa Milano, to increase awareness of our widespread sexual harassment problem. As a result of #MeToo, women—and men—around the country went public with their own stories, and that led to the downfall of actor Kevin Spacey, then Minnesota senator Al Franken, casino billionaire Steve Wynn, venture capitalist Dave McClure, comedian Louis C.K., celebrity chef Mario Batali, and even former US president George H. W. Bush. The 2017 *Time* magazine Person of the Year was "The Silence Breakers," a group of women who were brave enough to speak out.[25]

In the US workplace, 71 percent of employees report having been sexually harassed, compared to 40 percent in the United Kingdom[26] and about 35 percent in the Asia-Pacific region.[27] Even though Japanese women have been guaranteed equal opportunity in the workplace for over thirty years, the president of NH Foods Ltd., a Japanese food-processing conglomerate, had to step down after his subordinate made sexually explicit remarks to an airline employee while traveling. #MeToo affects everyone, everywhere. Yet only about a fourth of victims report their incidents to HR.[28]

While the majority of big-name harassment and assault allegations have been leveled by women against men, as I mentioned previously, sexual harassment is about power and domination. Powerful women can—and do—harass. For example, California assemblywoman Cristina Garcia (who, ironically, was one of the women featured in the *Time* magazine article) was accused by multiple male staffers of harassing them.[29] And just think of the many female high-school teachers who have been arrested for having sex with teenage male students.

The #MeToo movement has impacted the workplace in several ways, not all of which are good. These include limiting alcohol at office parties, using love contracts (romantically involved coworkers have to sign an agreement that they're together voluntarily), and employees being afraid to hug one another. I've also heard of cases of men shying away from networking with, mentoring—or even being alone with—female employees. That's tragic. At Facebook and Google, employees are allowed to ask out a coworker only once. "I'm busy" or "I can't that night," counts as a "no," said Heidi Swartz, Facebook's global head of employment law.[30] While the #MeToo movement has given safety, voice, and power to women (and men) at work, a side effect has been a 4 percent decline in office relationships over the past decade. Since we spend so much of our adult lives at work, we depend on it as a natural source for finding a mate. While banning romance (or even touching) at work can hurt our relationships, health, and well-being, we need to be aware when unwanted harassment occurs and stop it in its tracks.

Becoming aware of sexual harassment starts by understanding the legal definition: "Any unwelcome sexual advance or request can constitute illegal harassment if it creates a hostile work environment."[31] Unfortunately that definition is frustratingly broad. In truth, sexual harassment is often a "you-know-it-when-you-feel-it" kind of thing. However, if you haven't experienced sexual harassment directly, you may lack the empathy to understand how a victim feels. So, here's a way to think about it: as a leader, you're in a position of power and have influence over your team, their salaries, and their career trajectories. You should use this privilege only to support them and not take advantage of them or demoralize them. There's a fine line between wanted and unwanted gestures. For instance, never share inappropriate images (especially anything remotely sexual), tell sexual jokes, or send suggestive emails. But don't worry about handshakes or lunch dates.

If you see something that you feel might constitute sexual harassment, you have a variety of options. John Huntsman, associate director of information and data management at Bristol-Myers Squibb, tailors his response to the individual situation. "Often I find that the victim today stands their ground and pushes back, so in these cases I let them do that and provide backing as needed," he says. However, "in moments where the victim seems to be vulnerable, I will sometimes provide that pushback for them and socially call out the offender, although I try to do [so] in a manner that keeps the lines of communication open and provides a path for them to make amends."

If you feel comfortable dealing with a situation of sexual harassment on your own, be sure you document everything that happened and every step you take. If you don't feel comfortable or you're not sure what to do, take the matter to HR. The same logic applies if an employee comes to you with an allegation of sexual harassment against someone else: You must either investigate immediately or turn the matter over to HR. Failure to take allegations seriously has aggravated the problem and contributed to the creation of a toxic culture in which that type of behavior is acceptable, such as what existed at Fox News under former head Roger Ailes.

"Showing empathy as a leader starts well before the sexual harassment incident does," says Jenna Lebel, vice president of brand and integrated marketing at Liberty Mutual. "Empathy starts with creating an environment in which it's accepted and supported to step forward to report workplace harassment and discrimination."

While it's important to respond to an accuser's allegations, it's also important to respect the rights of the accused. "Making sure the allegations are 100% accurate is a whole other ball game," says Malcolm Manswell, marketing manager at Atlantic Records. "People have used this movement to tarnish the reputations of higher ups."

Vulnerability and Empathy

If you truly want to be an empathetic leader, you're going to have to do something that might be scary: show your own vulnerability. We're secretly envious of superheroes' powers in the movies, but it's their weaknesses that make them relatable and human. Our yellow sun gives Superman his powers, but kryptonite weakens him. If Superman had no weaknesses, watching him win every battle would be boring.

It's one thing to tell people about your talents and quite another to open up about your shortcomings. "Vulnerability is the birthplace of connection and the path to the feeling of worthiness," University of Houston Graduate College of Social Work research professor Brené Brown told me. "If it doesn't feel vulnerable, the sharing is probably not constructive."

Creating a Culture of Empathy: It's Up to You

Bad leadership and even worse behavior happen in both the physical and the online worlds, and the situation isn't getting any better. From a young age we learn that it's better to lead with our heads. But if we truly want to inspire and connect with others, we need to lead with our hearts, showing empathy and compassion to those around us. As a young leader, it's up to you to change this dynamic.

Instead of trying to quickly solve your employees' problems, you need to spend time listening to what's really going on with them and using your own experiences to better understand their emotions. Like me, millions of people have been bullied in their lives, and it has a devastating impact on our self-confidence in and out of the workplace. As a victim, I felt that anything I said out loud would be criticized, so I became quiet, extra careful with every word that came out of my mouth. It's taken me years to gain the courage to share the pain and trauma of my childhood bullying, but when I tell people what I've been through, they're often more open about their own pain, so I don't feel alone.

And try, as much as possible, to communicate less with technology. Instead of racing to collect hundreds of likes for a photograph, why not pick up the phone and tell someone how grateful you are that she has benefited your life? Instead of constantly watching your post to see who comments, why not invite someone for a coffee date?

Key Takeaways from Lead with Empathy

1. **Be vulnerable in your team conversations.** This humanizes you and makes it easier for your team members to approach you when they experience problems. Vulnerability isn't a weakness; it's a strength that creates a safe space and allows people to have a deeper relationship with you.
2. **Be fully present.** Listen to your teammates when they speak to you, setting aside any distractions (including your phone).
3. **Put others first.** When you focus too much on your own career, gaining power, and making money, you lose track of the people who can help you achieve all three. Putting yourself in their shoes will help you solve their problems or cater to their needs. This works better if you've endured the same tragedy or obstacles in the past, but even if you haven't, try your best to take a step back and think about it.

Chapter **10**

Improve Employee Experiences

You have to be the force that pumps information, drives communication, and maintains the culture across your teams.

—General Stanley McChrystal[1]

The word *experience* has become one of the defining business words of my generation because it takes into account every interaction you have with a person, place, product, or company. As a customer, the type of experience you have with a company will determine where you fall on the continuum between highly loyal, unpaid brand evangelist and toxic customer who goes to great lengths to slam the brand wherever and whenever possible. For employees, there's a similar continuum—one that ranges from loyal, productive workers who stay with you for years to disloyal, unproductive, destructive workers who undermine your team and your company. Where employees fall on that continuum is largely dictated by the employee experience you create for your team.

That experience is more complex than you might think, which is why I'm going to walk you through every aspect of it and how to improve it. Because it involves the physical, social, and cultural elements of your workplace as well as every touch point you have with your employees, creating a positive employee experience is going to take a lot of thought, plenty of creativity, and consistent effort.

In previous chapters, I have discussed many of the components that contribute to the employee experience. Now it's time to see how all those pieces work together. But rather than rehash them individually, I've come up with five rules to consider when thinking about the employee experience.

Rule 1. Be consistent in the way you treat your employees. If they see a colleague getting special treatment or having a unique experience with you, they'll feel left out and unappreciated.

Rule 2. Put effort into creating a culture that can be maintained even if you're not present. Don't expect your employees to blindly buy into your existing culture.

Rule 3. Strive to understand what makes them tick and how you can support them as individuals, not just as team members. Don't assume that your employees' needs are being met.

Rule 4. Empower employees to be part of the creation process and to foster the same experience for others. Don't try to take full responsibility for their experience.

Rule 5. Rely less on devices, platforms, and robotics. They remove the human touch that makes work personal and serves your employees' physiological needs. Don't trust technology to do the work for you.

Following these five rules will give you a better sense of what to avoid as well as where to focus your efforts. Remember, the employee experience you create is all encompassing and incorporates everything from the moment someone sends you a résumé all the way through their last day on the job. Let's take a closer look at the entire employee experience life cycle.

The Employee Experience Life Cycle

To create the best experience possible, you need to understand things from your employees' points of view, not just your own. There are six distinct periods in the employee experience life cycle that you need to pay attention to, covered in the following chart. In any major project with big milestones, different employees will need different amounts of your attention at different stages.

Employee Perspective	Employer Perspective	How to Handle
Joining	Recruiting	During interviews, help candidates understand your values and culture. Ask questions about the type of people they work best with and the experience they want to have at work every day. Think about how their personalities would mesh with the rest of the team, and if possible, let each team member meet each candidate before making a hiring decision. Most important, make sure you ask about their plans for the future. Before making a hiring decision, it's essential to make sure that the candidates' expectations align with the reality of the workplace and culture.
Familiarity	Onboarding	Get your new hires acquainted with the culture of the organization by arranging team lunches or meetings with smaller groups. Teach them the basics of their daily tasks and make sure they have the tools to get their daily work done. Assign a senior team member to mentor the newbies until they're able to stand on their own.
Learning	Development	Create a shared learning environment that encourages employees to help one another when they need it most. Have informal conversations with each employee. Besides making them feel special, that will also help you ascertain what their individual learning style is so you can maximize learning experiences and opportunities.

(continued)

Employee Perspective	Employer Perspective	How to Handle
Performing	Performance management	Keep tabs on how employees perform and intervene when necessary to ensure that they're always on the right track and feel confident about the work they're producing. Collect and distribute feedback regularly so they know where they stand, how to improve, and how to always be a solid team player.
Growing	Career advancement	After evaluating your employees, ensure that they have the necessary skills, leadership ability, and confidence to advance to the next level. Support and encourage them along the way. Understand their ambitions and what growth and advancement mean to them.
Leaving	Offboarding	You want departing employees to leave on a positive note and to speak highly of you, your team, and your company. And who knows, they may also return, or "boomerang," at some point in the future if they (and you) realize that you want to work together again.

What Employee Experience Is—and Isn't

The employee experience is the sum of all the interactions that affect employees' cognition, behaviors, and feelings. Their experience includes conversations with their teammates, the physical space they occupy every day, the nature of the work they do, and their observations throughout their journey at your company. It's how they feel about their workplace, their jobs, their teammates, and their bosses.

Creating the right employee experience is *not* about selecting a random assortment of perks, like Ping-Pong tables and free snacks, and sitting back in your office waiting for the magic to happen. Those perks sound cool, but they are addressing only short-term desires and don't play much of a role in the long-term employee life cycle. They don't engage your employees, help them become better at their jobs, or make them want to hang around at your company longer. Unfortunately, creating an optimal employee experience doesn't happen overnight. You must take a long, hard look at the entire life cycle and improve one facet at a time.

The Three Dimensions of Employee Experience: Culture, Relationships, and Space

When thinking about the different touch points we have with our teammates, we need to focus on three dimensions. You can control various levers in each dimension in a way that will affect how your employees feel, but over time things must be set up to run without your having to be involved. Let's take a detailed look at each dimension.

Culture

These are the unwritten rules of how teammates work together to accomplish goals and the glue that creates a cohesive, well-oiled team. Culture is made up of many elements, including core values, empathy, community, work ethic, language, symbols, systems, ethics, and rituals. It's the corporate version of a cult. When I was working at a big company, I'd use language that made no sense to my friends, parents, or even peers at other companies. Yes, that "secret" language made us feel a little cult-like, but it also brought everyone closer together because it was something exclusive that we all shared.

Stephanie Bixler, vice president of technology at Scholastic, told me about how one of her former bosses used the cult approach to motivate his team to succeed. "He decided to name his team GSD (Get Sh*t Done). He made us hats with this acronym and put up large signage around our team's space at work. He drilled the acronym into everyone's everyday language around the company [so] that it became its own brand at work," she says. "He positioned us as the elite, can-do group at the company; no problem was too difficult or big for us to solve. It gave us a sense of belonging and pride in what we were doing. Humans are wired for competition. And this type of branding and team building reinforces those basic instincts."

How important is culture? Researchers at the University of Southern California studied 759 firms in seventeen countries and found that the biggest driver of innovation wasn't salary or government policies; it was a strong corporate culture supported by the people who work there.[2]

Leaders who don't empower their employees—who act like taskmasters instead of orchestra conductors, who treat their employees like numbers instead of individuals—end up creating dysfunctional or failed cultures. When employees feel powerless to make decisions, don't receive feedback on their work, or aren't told how what they're working on fits into the bigger picture, they become less committed to achieving high standards of success. When departments don't communicate with one another or employees undermine their teammates, the organization begins to fail, and the culture becomes toxic.

Relationships

This is a critical part of employee experience because people connect emotionally with other humans much more than they do with a logo, a brand, or a product. If you treat your employees unfairly or you have a toxic employee who aggravates everyone else, the good people on your team are going to leave—and you shouldn't blame them. The best leaders and companies are the ones that create a family-like environment. They know that when you care about others' success, you create a strong emotional connection.

I had a conversation with Peter W. Schutz, the former CEO of Porsche, about his greatest challenge as a leader in the early 1980s, which was to "restore a dying organization, which was losing money, to growth and profitability." Faced with that problem, a lot of CEOs would have started by cutting costs, developing new products or services, inventing new marketing concepts, or coming up with clever advertising. But Schutz decided to start by rebuilding the culture, putting his employees' experience first. He believed that if all his employees—from the mailroom staff to the engineers and salespeople—felt like family and strove for shared success, Porsche would improve the quality of its products and start winning major races again.[3] He was right. With better engines, racing success followed, and worldwide sales grew from twenty-eight thousand to fifty-three thousand units per year from 1980 to 1986.[4] Corporate profitability increased as well.

Leaving a job is easy, but leaving a family is a lot harder. My managers at the two corporate jobs I had immediately before I started my own company cried when I transitioned.

Employees who feel like family are likely to socialize with one another, and that promotes and improves teamwork. In one study of twenty thousand employees, researchers found that those who knew three or more people at their company were likely to stay there longer.[5] A lot of this socializing and relationship building is your responsibility. In a separate study, the same researchers gathered data from fourteen hundred supervisors and thirty thousand employees and discovered that an employee's first manager had the biggest impact on their performance years later.[6] By maintaining a strong relationship with your employees, you can see their performance accelerate every year. I talk a bit more about your role in relationship building below.

Space

Space is your employees' physical environment, the place where their senses are being put to work touching, tasting, seeing, and smelling everything from the cafeteria food to their office environment to the holiday decorations. The age of the people they work with, the office layout, and the lighting all matter to employees, even if they never mention these things. Physical space is key to creativity, collaboration, and wellness at work. If you don't get space right, another company will. Employees want to be comfortable and want their individual work preferences accounted for. One employee may prefer a cubicle, whereas another might prefer to work in a lounge, and these preferences can change regularly. At DELL EMC, leaders worked with their building teams to modernize their office spaces with new technologies, Adam Miller, product marketing manager at the company, told me. This included "standing desk options, informal meeting spaces, and more. These enhancements gave employees more flexibility to work in ways that are productive for them." And at Cisco, the new CEO allowed employees to bring their dogs to work, says Caroline Guenther, integrated business planning manager at the company. Giving your employees options lets them select the environment that will make them more productive and more creative.

Despite the increasingly mobile—and remote—workforce, space still plays a key role in how we experience culture, cultivate relationships, and solve business issues. Researchers Craig Knight and Alex Haslam gave forty-seven office workers in London the option to arrange their office with as many plants and pictures as they wanted.[7] Those workers were 32 percent more productive and were more committed to their team's success than a control group of workers who didn't get to decorate their offices. In a different study, about half of those polled said that an office redesign would increase productivity, make them better organized, and increase their job satisfaction levels.[8] And a study by the American Society of Interior Designers found employees who like their office environment are 31 percent more likely to be satisfied with their jobs.[9]

On the other hand, employees who are bombarded with loud noise throughout the day, have bad lighting or air quality, or work in a technologically outdated office or in a building that's isolated from parks or pleasant outdoor spaces are less likely to be excited about their jobs, less willing to work longer hours, and less likely to produce for you. Space influences our mood, behaviors, and overall impression of whom we work for.

When it comes to space, we need to provide flexibility and options and encourage our employees to be honest about how we can improve it. And since you're wondering, I'm still not talking about Ping-Pong tables, free snacks, and brightly colored slides between floors. Those are accessories that can make a good thing better, but they won't be enough to keep employees from jumping ship if their overall physical environment is poor.

Space enforces and reinforces your culture every single day. Although your company controls the light switches and layout, employees should be able to decide how to personalize their cubicles or offices. If you're nice about it, you may be able to influence them to make changes (e.g., you might point out that their productivity and health could be negatively affected by having a cluttered desk or nonergonomic screen or keyboard placement), but ultimately, it's up to them to make decisions for their own benefit.

Let Your Employees Define the Experience

Instead of creating an employee experience in isolation or from the top down, why not encourage and empower your employees to be part of the creation process? If you let your team give you feedback and share ideas about the experience they desire, it'll be much easier to meet those expectations. For Erin Yang, vice president of technology product management at Workday, one of the ways her team members' experience was improved was by giving them the opportunity to participate in defining it. "I was nominated to be on a steering committee that helped design a new floor in our San Francisco office," she said. "We were able to customize the floor to be optimal for the way product management and development teams worked. I in turn looped in my own team by asking them to contribute office ideas on a shared Pinterest board. This made us much more engaged in the new space that was created."

Aside from letting your employees define their workspace, let them have a hand in other areas of their work experience. Erin told me that she sees this happening in other aspects of Workday, such as their snack program. "Our Employee Programs group regularly polls employees on what snacks they would like, and importantly, they actually make changes based on the feedback. This is something I see that people appreciate."

The most effective way to empower your teammates to help create an experience that meets their needs and expectations is to simply give them a seat at the table. Make sure they know that their opinions and thoughts impact change because, as I mentioned in chapter 1, people intrinsically have to feel that their work matters. Regardless of their title or tenure, including your employees in important discussions makes them feel important, while it simultaneously ensures diverse ideas.

"In moments where I haven't been confident enough to ask for a seat at the table, my director pulls out a chair for me—both literally and metaphorically—and invites me to sit down," says Katie Lucas, senior manager of digital content at HBO. "He looks for opportunities to elevate me and empower my work." By giving your teammates a (literal or metaphorical) seat at the table, you're involving them in the decisions that will ultimately influence their employee experience.

How Employee Experience Makes a Business Difference

As with anything that affects the workplace, we need to be able to justify focusing on improving the employee experience. Fortunately we've been able to measure the business benefits companies receive when they give their employees a positive, memorable experience throughout their life cycle: they'll stay with you longer, perform better while they're there, and serve as unofficial brand ambassadors who can help with your recruiting. When we surveyed executives, more than 80 percent said that employee experience is either important or very important to their organization's success, versus a mere 1 percent who said it isn't important.[10] I predict that 1 percent will be looking for new jobs soon. In a separate study, Deloitte not only confirmed our data but also discovered that only 22 percent of companies were excellent at building a differentiated employee experience.[11]

IBM and Globoforce have been able to link various aspects of work to an employee's overall experience.[12] They found that positive experiences are associated with better performance, lower turnover, higher levels of social relatedness, and better teamwork. For instance, when employees feel their ideas are heard, more than 80 percent report a more positive experience.

The following exercise will help you identify areas in which you can improve the employee experience. These are based on the feedback you've received, the daily behavior of your teammates, and the actions that have or haven't been taken to make the workplace better.

Self-Reflection: What Experience Are You Creating for Your Employees?

To effectively review and improve your employee experience, ask yourself the following questions, which will give you a better sense of what you've done (or haven't done) to ensure that your employees are satisfied and engaged. If you have trouble answering these questions, or if you need additional data, I highly recommend that you set up one-on-one conversations with your employees and ask them how they view their experience. Your goal should be to at least meet (if not exceed!) their expectations.

1. What are the top three things that you believe drive your employees?
2. Have you noticed team members socializing both in and out of work?
3. Do your employees know the mission and purpose of your company?
4. What feedback have you received from job seekers or former employees about their experiences?
5. Is your employee turnover too high?
6. How much thought are you putting into hiring for personality fit?
7. Have you examined your office space's impact on employee productivity?
8. Are employees receiving the proper amount of support to complete tasks?

9. Have you created a family environment in which employees feel safe?
10. What aspects of the employee experience life cycle can you improve?

How to Improve the Employee Experience

Now that I've covered why employee experience is so critical to your company's success and you know where you need to improve, it's time to talk about some key strategies. In our research, HR leaders from around the country told us that the top three ways to enhance employee experience are to

1. invest in training and development,
2. improve employees' workspace, and
3. give more recognition.[13]

This makes perfect sense, right? When employees have the necessary courses and resources to hone their skills, they're going to be more confident in meetings and more likely to share their knowledge with the team. The workspace, as already discussed, is crucial because we spend so much time there. And recognition makes employees feel good about themselves and creates a culture in which they see and appreciate the positive qualities and accomplishments of others.

Trying to improve your employee experience without having measured it first is a waste of time. To get the most accurate measurements of the experience itself, you need to see how well it matches the expectations of people who currently work for you or want to.

Outside your company, there's a lot of information available from review websites and professional networks where you can get uncensored and anonymous feedback concerning how job seekers and employees have been treated. Every weak point you identify is an opportunity to close a gap. For instance, if an employee feels that what they were hired to do didn't turn out to be what they'd actually been doing and they quit, you clearly need to update your job descriptions and tweak your onboarding. If you see multiple comments from former employees about how management didn't support them, you might need to take a course yourself or alter the management-training curriculum you run your team through. This feedback is quite common because employees leave poor management more than they leave jobs.

Inside your company, look at every aspect of the employee experience life cycle from when employees join to when they eventually leave. The easiest way to get the most current information is to use your team as a focus group at least once a month and ask them for their candid feedback on how to improve the work environment. "We did a wonderful exercise where people could put Post-its up in designated spaces in the office with feedback on changes they'd like to see," says Sarah Unger, vice president of marketing strategy, trends, and insights at Viacom. "But the key thing is—the changes were actually listened to."

You should also collect information from job candidates during onboarding and from employees who are leaving during offboarding. Together, these before, during, and after data points will give you a more complete picture of the experience they've had. If someone started their job with great enthusiasm but ended it with frustration, you need to find out why so it doesn't happen again. Compare happy and unhappy candidates and employees to understand where the gaps are and how to fill them.

Your Role in Creating Employee Experience

You can have the most amazing workspace, but without strong relationships between management and employees, you will fail. Guaranteed. That's why you need to both become a better leader and encourage or train others to do the same. Managers have an astronomical impact on their employees' experience because they're constantly interacting with them, from asking for advice to assigning new projects. By having a transformational leadership style, being open to—and welcoming—feedback, and encouraging the best in everyone, you'll be more appealing to them, and they'll work harder for you.

I've heard people say that good managers are born, not made, but in my experience good management can be taught to anyone who's willing to put in the time to learn the key skills required to do it. We need to stop promoting managers simply because they have tenure or work hard or because we're afraid of losing them. Giving mediocre or poor managers even more power is an excellent way to undermine the employee experience.

When it comes to creating a powerful and exciting experience for your employees, seemingly trivial things matter. One of the best ways to foster strong team relationships is to think outside the corporate walls and plan social outings and events for your employees. Dinners to show your appreciation, celebrate a birthday, or highlight a milestone can really bring your team together. Sadly, most companies are so focused on the bottom line that they don't see strong employee relationships as a key component for generating higher revenue. In one study, placement giant Robert Half found that 80 percent of companies don't hold annual gatherings.[14] A simple, public "thank-you," a party, a gift card, tickets to a sporting event, or a free dinner can go a long way. You may not think it's a big deal, but to your employees it is—especially when they don't expect it.

When your employees see that you support outside-the-office socialization, they'll start to do the same. If they don't, openly encourage them to do so. When employees behave in a friendly way toward others on their team, they're creating a community that they'll want to stay in for the long haul. When communities have a strong social bond, in which employees trust one another and care about one another beyond simply finishing the latest project, companies thrive. Strong empathetic gestures, such as visiting a team member who's in the hospital, have a major impact on employees' livelihood and help them see you not only as their manager, but as a friend who genuinely cares about them. When coworkers are playing on a company sports team or are part of a resource group to help other employees (for women, young professionals, Latinos, etc.), you know that they're building and growing their relationships—and their team—without having been forced.

Keep in mind that outside-the-office socializing is only one of many components of the overall employee experience. "Fun, social stuff is amazing and funny but when bosses rely on that over tone of the work environment it feels like the wrong priority to me," says Amanda Pacitti, VP of learning and development at Time, Inc. "I don't think anyone has ever said, 'Wow, my boss took me bowling, I love my job!' ever." Amanda Zaydman, brand manager at Nestle Purina, adds that "people want to be inspired, feel valued, and believe in the work they're doing. Sometimes that means taking the time to throw [a] surprise baby shower or pick up cupcakes for a birthday, but I think what you do day to day has a greater impact. Being honest and transparent. Listening. Advocating for talented employees to get the assignments and promotions they want and deserve. It's not anything revolutionary, but it's easy to forget as a manager."

Aside from celebrating with your teammates, you want to get to know them on a personal level. We may all have similar basic needs and desires, but as individuals we also have unique motivations and dreams. As you're getting to know each of your employees, write down their biggest drivers, interests, and aspirations so that you can consciously work toward meeting their needs.

Face time with leaders can play a critical role in influencing the workplace experience. Lindsay Weddle, land manager at ConocoPhillips, had an experience with a leader in her company whom she greatly admired but had never worked for. What he did influenced her managerial style later. "At some point during a meeting, I told him my daughter's name. Months later he walked into an elevator and said, 'Hey! How's Abigail?' Quite frankly, I was stunned. I was so impressed he remembered something personal about me and took the time to talk to me about my daughter," she told me. "He was no doubt one of the busiest people in the company, but that moment of thoughtfulness has stuck with me years later. It is the reason I make a conscious effort to memorize spouse and kids' names and ask by name how they are doing."

The following table will help you support employees, who each have their own set of motivations and interests at work.

Employee Motivations and Interests	How to Support Them
Flexibility	Give employees permission to come into work later or telecommute at least one day per week.
Pay	As long as they're performing well and driving business results, make sure they get a pay increase at least once a year.
Friendships	Introduce them to people inside and outside your team and invite them to social events.
Sports	In addition to a pay bonus, give them two tickets to a sporting event you know they'll love.
Travel	If you have offices in multiple cities, allow them to work from one of those offices. Or if there's an industry conference, let them attend so they can learn and travel simultaneously.

Getting to know your employees isn't a one-way street. "I've improved the experience employees have at work by getting to know them on a personal level, and allowing them to know me on this same level in return," says Amanda Healy, senior marketing manager at TIBCO Software. "Being a different person at work versus at home is exhausting—for me, what you see is what you get. I'll talk about my husband, I'll send along a song I'm jamming to, I'll share photos of my cats. Personal details are the connective tissue that drives myself, and my business, forward."

When it comes to building positive employee experience, the biggest mistake I see managers make is to *not* set realistic expectations during new hires' first ninety days. It's important to let them know what they'll be learning, establish goals, and set the agenda for their job responsibilities moving forward. Make them feel it's not just a job but part of their long-term careers. No one wants to feel like a robot or a mindless cog on an assembly line; people want to know how their jobs factor into making the company successful. "Key for me is making people understand that their contribution is appreciated, and it makes a difference, whether it's people working for me or not," says John Mwangi, vice president of information governance, law, and franchise integrity at Mastercard. "Knowing that your work matters significantly improves a person's experience and they view you as a partner in their professional development."

Map out employees' training plans with them, including specifics on the skills they'll need to move ahead, suggested courses to complete, and an explanation of how this will help them perform better. Doing this will help you reduce the stress and anxiety that people naturally have when they start something new.

As much as you'd like to control your employees' experience, you can't—at least not all the time. Just as with consumerism, in which people have some control over the brand through word of mouth, your employees can either talk up or talk down their experience to others. That's why it's so important to empower your employees to own their experience by providing them the right tools and a dedicated support system. The more autonomy you give them (assuming they can handle the responsibility), the less pressure there will be on you.

Vivek Raval, head of performance management at Facebook, does a nice job of summing it all up. "The best leaders I have worked with took a personal interest in my development and growth," he says. "They asked about my objectives, offered new ideas on how I could develop and grow, spent time to understand my working style, and most importantly took actions throughout their time with me to put me in positions to learn and succeed."

Key Takeaways from Improve Employee Experiences

1. **Focus on the most critical experiences.** Think of the entire employee experience life cycle and try to improve one part of it at a time so you won't be overloaded. Aside from your everyday interactions with your team members, focus on the most critical periods of their experience from their first day on the job to their last.
2. **View the employee experience from their perspective.** Use internal or external data to identify areas of improvement. Turn your team into a focus group and encourage honesty by being transparent with them about how you feel and the feedback you've received from others.
3. **Empower your employees.** Provide employees with some degree of control over their own experience by giving them the freedom to chart their own paths. Get to know them on a personal level so you can help them with their individual development and support their ambitions.

Conclusion

Become More Human

Don't confuse movement with progress.

—DR. MEHMET OZ[1]

As technology emerges, advances, and revolutionizes every industry, profession, and culture, we've barely scratched the surface of how devices, networks, and artificial intelligence will change human behavior; displace jobs; and impact our organizations, communities, and lives. We thought that technology would bring us closer together, yet it has made our work lives more challenging and less meaningful. In the not-too-distant future, robots may bring you your morning coffee and brush your teeth for you, but you'll still have a heart, a soul, and a mind—and so will the people who work for, and with, you. The essential qualities that you'll need to be an effective leader, such as empathy, openness, and vision, can't be outsourced to machines. For that reason, as leaders we need to get back to human and become the masters of technology instead of the other way around.

A Warning About Technology from Prominent Technology Leaders

There's no stopping the changes brought on by the technology revolution, but we must embrace them with caution. That's a view I share with many of the most respected experts in technology and artificial intelligence. And when they warn us, we really should take heed. Steve Wozniak, Stephen Hawking, and Elon Musk, for example, signed an open letter on the societal impacts of artificial intelligence (AI).[2] And Microsoft's research director, Eric Horvitz, believes that someday AI could turn against us and become a threat to our very existence.

Two other tech luminaries, Apple CEO Tim Cook and Facebook CEO Mark Zuckerberg, gave commencement speeches warning of the pitfalls of the tools and systems they themselves are promoting. At MIT, Cook said, "Sometimes the very technology that is meant to connect us, divides us. Technology is capable of doing great things. But it doesn't want to do great things."[3] At Harvard, Zuckerberg said, "When our parents graduated, purpose reliably came from your job, your church, your community. But today, technology and automation are eliminating many jobs. Membership in communities is declining. Many people feel disconnected and depressed, and are trying to fill a void."[4]

Being a young leader in the tech-saturated world we live in is a challenge. And we'll be successful only if we're able to create emotional connections to others—the type of connections that enable us to empathize, perform acts of kindness, and avoid hurting others.

Workplaces Are Increasingly More Robotic

In the Virgin Pulse study, we asked employees and managers which trends they believe will most impact their work experience.[5] They said they are paying most attention to the Internet of Things, AI, advancements in smartphones, virtual reality, and wearable tech—in other words, automation. How big an impact will automation have? We surveyed hundreds of organizations and found that on average they expect to reduce their workforce by at least 10 percent over the next few years.[6]

A lot of people think of robots as something futuristic, but we're already a lot closer than you might imagine. McDonalds is replacing cashiers with kiosks,[7] Domino's Pizza is replacing its delivery teams with self-driving robots,[8] Lowe's has replaced human greeters with robo-greeters,[9] and Aloft Hotel is experimenting with robotic bellhops.[10] No job is safe. In China, legal robots have been deployed to decide sentencing in certain court cases; it doesn't get scarier than that.[11] Bottom line: automation will completely eliminate a swarm of jobs from our global economy as well as a variety of tasks within jobs that it doesn't kill.

From the company perspective, robots are a way to lower labor costs and increase profitability. Think about it. If a company can purchase a robot for a one time investment of $30,000 instead of hiring a full-time employee to do the same dozen tasks for $75,000—plus health care, paid vacation, and the potential for raises and bonuses—it'll choose the robot. The robot can work twenty-four-hour days, whereas the human might max out at eight. The robot won't argue with you about your process or complain that it's burned out or stressed; it'll do what you tell it to with no complaint. As the cost of these machines inevitably declines, they'll become even more attractive to employers, which is exactly what's going on in the minds of many CEOs around the world.

There's no doubt in my mind that technology continues to disconnect us from other humans—even as it becomes more personal, whether it's virtual reality, chatbots, or microchip injections, which sound like something from the latest sci-fi movie but are already happening. Epicenter, a Swedish firm, has offered to inject employees with microchips at no cost and already has 150 takers. Although employees with the implanted chip can easily access doors and office utilities like photocopiers without having to take out their wallets, they're constantly being tracked, which sounds extremely invasive to me. Imagine having to get that chip surgically removed if you want to switch employers!

It's Time to Be More Human and Less Machine

I have not only witnessed, but also participated in, the back-to-human renaissance. When technology makes me feel isolated from others, I naturally feel the need to connect more. Whether it's meeting someone for coffee, walking to the office, or even calling my parents, I try not to let technology get the best of me. Instead, I use it to create more in-person situations and deeper conversations with others. In the workplace, our teams can't function if we aren't there to support them, and without a sense of connection, they won't be as committed or productive. While machines become adept at hard skills and perform many tasks more quickly than humans will ever be able to, humans will always have the upper hand when it comes to the soft skills that make great leaders.

In an interview with CNBC, Sinovation Ventures founder Kai-Fu Lee was asked whether humans still have a place in a world where machines are growing more intelligent. Even though he invests in technology, Lee admitted, "Touching one's heart with your heart is something that machines, I believe, will never be good at."[12] Today's jobs emphasize the ability to work with or for other people much more than they did in the 1980s or 1990s.[13] As jobs are lost, new ones are created, and those will continue to require leadership, teamwork, time management, and social skills. Your ability to develop strong work relationships will be your most important asset as you become the leader you aspire to be. Despite our technological future, our social skills will be the fabric out of which we'll continue to weave our careers and lives. As *Fortune Magazine* senior editor Geoff Colvin told me, "We're hardwired by 100,000 years of evolution to value deep interaction with other humans and not with computers."[14]

We must start by acknowledging that we need to use technology to foster deeper connection and stronger relationships. And we must continue by admitting that we need not just more friends, but deeper conversations with our current friends. And I don't mean those superficial "friendships" you have online, where you look at updates, likes, and comments but rarely if ever get on the phone with them or even wish them a happy birthday. I mean the friends that you invest your time in, the ones you genuinely care about, and the relationships you have with the coworkers you see or don't see every day.

Everything we need to succeed, from learning to emotional support, can be improved through having friendships. There's a reason that phrases like "Your network is your net worth" and "It's not what you know, it's *who* you know" get passed down from one generation to the next. They're true! People, not machines, are going to lead you to knowledge, jobs, and fulfillment. Over the years I've made it a point to ask older adults about their friendships, and every one of them has told me what researchers have been saying for years: as you get older, you have fewer close friends. Several studies have also concluded that people regret relationship mistakes more intensely than any career decisions.[15] Knowing this can help you decide whom you hold close and whom you let go. As we get older, we have more responsibilities, from bearing children to overscheduling, and our friends get shortchanged as a result. You can do something about this and create a more fulfilling work experience for your employees in the process.

The Future Is Now

In the age of isolation, there is light. Today, I challenge you to put your phone down, turn off your notifications, and get offline. I know you can do it! There's no going back in time, but there *is* going back to human. So, make every interaction count every day, every hour, and every minute. I trust you to lead the way—and I'll be right there beside you to lend a hand.

Acknowledgments

To My Literary Agent

This book is dedicated to Jim Levine, my literary agent and hero. He has always believed in me and has had a profound influence on my entire career since we've worked together. Jim is an unsung hero in the publishing world, having been a driving force behind some of the biggest authors and ideas of our time, yet being humble and staying behind the scenes. He is an inspiration to me because he could easily retire on the golf course, yet he chooses to continue to help the next generation of authors succeed. Only someone who truly loves what he does, like Jim, could continue to both lead an admired company and manage an endless roster of authors. Although he calls me "scrappy," dedicating this book to him will motivate me to make it my best one yet and to succeed far beyond what even he had imagined.

To My Parents

Thank you for believing in me and being my sounding board. I love you both very much!

To Da Capo Press

Dan Ambrosio, John Radziewicz, Kerry Rubenstein, Kevin Hanover, Michael Clark, Michael Giarratano, Miriam Riad, and their team believed in the book concept and helped me bring it to life through words. Thank you for your vote of confidence, investment, and time.

To Armin Brott

Your editing has helped me become a better writer and has improved the quality of my books. I can't thank you enough for your effort, support, and encouragement.

To Professor Kevin Rockmann

Kevin signed on to develop the Work Connectivity Index (WCI) assessment for this book immediately, and it is my hope that his academic research on isolation will continue.

To Future Workplace

I appreciate the support of my team, including David Milo, Jeanne Meister, Kevin Mulcahy, Lea Deutsch, Tracy Pugh, and Tuan Doan. With their help we are influencing the next generation of leaders and making a difference in the workplace.

To Virgin Pulse

From my very first call with Wendy Werve, I knew Virgin Pulse would be the perfect partner for the book's global research study. A special thank you to Andrew Boyd, Arthur Gehring, Elise Meyer, and Hailey McDonald.

To CA Technologies

My professional speaking career launched when CA hired me to speak many years ago, and now things have come full circle as they host my national book launch. A special thank you to Laura Drake and Patricia Rollins for believing in me.

To the Millennial 100

For the book I interviewed a hundred of the top millennial leaders (The M100) from some of the most notable companies in the world. These include Adam Miller, Alison Elworthy, Amanda Fraga, Amanda Healy, Amanda Pacitti, Amanda Zaydman, Amit Trivedi, Amy Linda, Amy Odell, Andrew Miele, Antonio McBroom, Ben Thompson, Bill Connolly, Bill Wells, Bradford Charles Wilkins, Brandon Gross, Bryan Taylor, Carly Charlson, Caroline Guenther, Charlie Cole, Chris Gumiela, Dan Klamm, Danny Gaynor, Daniel Jeydel, Daniel Kim, Daniel LaCross, Danielle Buckley, Derek Baltuskonis, Derek Thompson, Ed Mendrala, Emily Kaplan, Erin

Millard, Erin Yang, Felipe Navarro, Heather Samp, Ilona Jurkiewicz, Jason Gong, Jenna Lebel, Jenna Vassallo, Jennifer Cochrane, Jennifer Fleiss, Jennifer Grayeb, Jennifer Lopez, Jennifer Schopfer, Jessica Goldberg, Jessica Latimer, Jessica Roberts, Jill Zakrzewski, Joe Lawrence, John Huntsman, John Mwangi, Justin Birenbaum, Justin Orkin, Kate Mangiaratti, Katie Lucas, Katie Vachon, Kiah Erlich, Kristy Tillman, Kyle York, Lara Hogan, Laura Enoch, Laura Petti, Leor Radbil, Lindsay Weddle, Liz Miersch, Malcolm Manswell, Mathew Mehrotra, Meg Paintal, Meghan Grady, Melanie Chase, Michelle Odland, Mike Maxwell, Mike Schneller, Dr. Nandi J. Shareef, Nawal Fakhoury, Nim De Swardt, Om Marwah, Paolo Mottola, Patricia Rollins, Paul Reich, Philip Krim, Rajiv Kumar, Rashida Hodge, Rosie Perez, Ross Feinberg, Sam Howe, Sam Violette, Sam Worobec, Sarah D'Angelo, Sarah Unger, Sarah Welsford, Sharmi Gandhi, Simon Bouchez, Sjoerd Gehring, Stephanie Bixler, Stephanie Busch, Tracy Shepard-Rashkin, Ulrich Kadow, Vicki Ng, and Vivek Raval.

To Three Bridges Productions

Thank you for creating a comedy sketch to illustrate the book's main message. The team included Alec Lawless, Catherine Beckett, Christian Roberts, Gina Ferranti, James Meeg, Lara Goldie, Peter Getz, Trevor Livingston, and Yasmeen Jawhar.

To My Friends

Throughout the writing process my friends not only supported me but were a great sounding board for some of my ideas. A special thank you to Allen Gannett, Chris Schembra, Cory Bradburn, David Homan, Farnoosh Torabi, James Altucher, Jay Shetty, Jeff Gabel, Jennifer Sutton, Jenny Blake, Jess Cording, Joe Crossett, Jonathan Mitman, Jordan Harbinger, Josh White, J. R. Rothstein, Julia Levy, Julie Billings-Nguyen, Labe Eden, Lewis Howes, Mickey Penzer, Mike Smith, Pete Ziegler, Rachel Tuhro, Russell Wyner, Ryan Paugh, Shane Snow, and Yoni Frenkel.

Notes

Introduction: How Technology Is Isolating Us at Work

1. Dan Schawbel, "Arianna Huffington: Why Entrepreneurs Should Embrace the Third Metric," *Forbes*, March 25, 2014, https://www.forbes.com/sites/danschawbel/2014/03/25/arianna-huffington/.
2. Jeffrey M. Jones, "In U.S., Telecommuting for Work Climbs to 37%," Gallup, August 19, 2015, http://news.gallup.com/poll/184649/telecommuting-work-climbs.aspx.
3. James Manyika et al., "Harnessing Automation for a Future That Works," McKinsey & Company, January 1, 2017, https://www.mckinsey.com/featured-insights/digital-disruption/harnessing-automation-for-a-future-that-works.
4. Rob Cross et al., "Collaborative Overload," *Harvard Business Review*, December 20, 2016, https://hbr.org/2016/01/collaborative-overload.
5. "How Americans Spend Their Money," *New York Times*, February 10, 2008, http://archive.nytimes.com/www.nytimes.com/imagepages/2008/02/10/opinion/10op.graphic.ready.html.
6. "Why Can't We Put Down Our Smartphones?," *60 Minutes*, CBS Interactive, April 7, 2017, https://www.cbsnews.com/news/why-cant-we-put-down-our-smartphones-60-minutes/.
7. Gavin Francis, "Irresistible: Why We Can't Stop Checking, Scrolling, Clicking and Watching—Review," *Guardian News and Media*, February 26, 2017, https://www.theguardian.com/books/2017/feb/26/irresistible-why-cant-stop-checking-scrolling-clicking-adam-alter-review-internet-addiction.
8. Sarah Perez, "U.S. Consumers Now Spend 5 Hours per Day on Mobile Devices," TechCrunch, March 3, 2017, https://techcrunch.com/2017/03/03/u-s-consumers-now-spend-5-hours-per-day-on-mobile-devices/.
9. Patrick Nelson, "We Touch Our Phones 2,617 Times a Day, Says Study," Network World, July 7, 2016, https://www.networkworld.com/article/3092446/smartphones/we-touch-our-phones-2617-times-a-day-says-study.html.
10. Eric Barker, "This Is How to Stop Checking Your Phone: 5 Secrets from Research," *Barking Up the Wrong Tree*, March 5, 2017, https://www.bakadesuyo.com/2017/03/how-to-stop-checking-your-phone/.
11. Anderson Cooper, "What Is 'Brain Hacking'? Tech Insiders on Why You Should Care," *60 Minutes*, CBS Interactive, April 9, 2017, https://www.cbsnews.com/news/brain-hacking-tech-insiders-60-minutes/.

12. "Despite the Tech Revolution, Gen Z and Millennials Crave in-Person Collaboration." Future Workplace and Randstad, September 6, 2016, https://www.randstadusa.com/about/news/despite-the-tech-revolution-gen-z-and-millennials-crave-in-person-collaboration/.
13. Miller McPherson et al., "Social Isolation in America: Changes in Core Discussion Networks over Two Decades," *American Sociological Review* 71, no. 3 (June 1, 2006): 353–375, https://doi.org/10.1177/000312240607100301; and Vivek Murthy, "Work and the Loneliness Epidemic," *Harvard Business Review*, September 27, 2017, https://hbr.org/cover-story/2017/09/work-and-the-loneliness-epidemic.
14. Dan Schawbel, "Vivek Murthy: How to Solve the Work Loneliness Epidemic," *Forbes*, October 7, 2017, https://www.forbes.com/sites/danschawbel/2017/10/07/vivek-murthy-how-to-solve-the-work-loneliness-epidemic-at-work/#c4bc48d71727.
15. Carolyn Gregoire, "The 75-Year Study That Found the Secrets to a Fulfilling Life," *Huffington Post*, August 11, 2013, https://www.huffingtonpost.com/2013/08/11/how-this-harvard-psycholo_n_3727229.html.
16. Hakan Ozcelik and Sigal Barsade, "Work Loneliness and Employee Performance," n.d., https://journals.aom.org/doi/abs/10.5465/ambpp.2011.65869714.
17. Stephen Jaros, "Meyer and Allen Model of Organizational Commitment: Measurement Issues," *ICFAI Journal of Organizational Behavior* 6 (November 4, 2007): 1–20.
18. Kerry Hannon, "People with Pals at Work More Satisfied, Productive," *USA Today*, August 13, 2006, http://usatoday30.usatoday.com/money/books/reviews/2006-08-13-vital-friends_x.htm.
19. Future Workplace and Virgin Pulse, "The Work Connectivity Study," to be published November 13, 2018, at http://workplacetrends.com/the-work-connectivity-study/.
20. Future Workplace and Polycom, "The Human Face of Remote Working Study," March 21, 2017, http://workplacetrends.com/the-human-face-of-remote-working-study/.
21. "Japan Population to Shrink by One-Third by 2060," *BBC News*, January 30, 2012, http://www.bbc.com/news/world-asia-16787538.
22. Alanna Petroff and Océane Cornevin, "France Gives Workers 'Right to Disconnect' from Office Email," *CNNMoney*, Cable News Network, January 2, 2017, http://money.cnn.com/2017/01/02/technology/france-office-email-workers-law/index.html.
23. "PM Commits to Government-Wide Drive to Tackle Loneliness," Gov.uk, January 17, 2018, https://www.gov.uk/government/news/pm-commits-to-government-wide-drive-to-tackle-loneliness.
24. Uptin Saiidi, "Millennials: Forget Material Things, Help Us Take Selfies," CNBC, May 5, 2016, https://www.cnbc.com/2016/05/05/millennials-are-prioritizing-experiences-over-stuff.html.
25. "Table 1: Time Spent in Primary Activities and Percent of the Civilian Population Engaging in Each Activity, Averages per Day by Sex, 2016 Annual Averages," US Bureau of Labor Statistics, June 27, 2017, https://www.bls.gov/news.release/atus.t01.htm.

26. J. Holt-Lunstad, T. B. Smith, and J. B. Layton, "Social Relationships and Mortality Risk: A Meta-Analytic Review," *PLoS Med* 7, no. 7 (2010): e1000316, https://doi.org/10.1371/journal.pmed.1000316.
27. Lydia Saad, "The '40-Hour' Workweek Is Actually Longer—by Seven Hours," Gallup, August 29, 2014, http://news.gallup.com/poll/175286/hour-work week-actually-longer-seven-hours.aspx.

Chapter 1: Focus on Fulfillment

1. Dan Schawbel, "Michael Bloomberg: From Billionaire Businessman to Climate Change Activist," *Forbes*, May 30, 2017, https://www.forbes.com/sites/danschawbel/2017/05/30/michael-bloomberg-from-billionaire-businessman-to-climate-change-activist/#1e00ede25a20.
2. Michael Bond, "How Extreme Isolation Warps the Mind," BBC Future, May 14, 2014, http://www.bbc.com/future/story/20140514-how-extreme-isolation-warps-minds.
3. Erica Goode, "Solitary Confinement: Punished for Life," *New York Times*, August 3, 2015, https://www.nytimes.com/2015/08/04/health/solitary-confinement-mental-illness.html.
4. Mark Molloy, "Too Much Social Media 'Increases Loneliness and Envy'—Study." *Telegraph*, March 6, 2017, https://www.telegraph.co.uk/technology/2017/03/06/much-social-media-increases-loneliness-envy-study/.
5. Melissa Carroll, "UH Study Links Facebook Use to Depressive Symptoms," University of Houston, August 6, 2017, http://www.uh.edu/news-events/stories/2015/April/040415FaceookStudy.php.
6. Mike Brown, "How Accurately Does Social Media Portray the Lives of Millennials?," LendEDU, May 15, 2017, https://lendedu.com/blog/accurately-social-media-portray-life-millennials/.
7. Holly B. Shakya and Nicholas A. A. Christakis, "A New, More Rigorous Study Confirms: The More You Use Facebook, the Worse You Feel," *Harvard Business Review*, April 10, 2017, https://hbr.org/2017/04/a-new-more-rigorous-study-confirms-the-more-you-use-facebook-the-worse-you-feel.
8. Future Workplace and Kronos, "The Employee Engagement Study," January 9, 2017, http://workplacetrends.com/the-employee-burnout-crisis-study/.
9. Staples Business Advantage, "The North American Workplace Survey," June 29, 2015, http://workplacetrends.com/north-american-workplace-survey/.
10. Jeffrey M. Jones, "In U.S., 40% Get Less Than Recommended Amount of Sleep," Gallup, December 19, 2013, http://news.gallup.com/poll/166553/less-recommended-amount-sleep.aspx.
11. "Overweight & Obesity Statistics," National Institute of Diabetes and Digestive and Kidney Diseases, US Department of Health and Human Services, August 1, 2017, https://www.niddk.nih.gov/health-information/health-statistics/overweight-obesity.
12. Millennial Branding and Randstad, "Gen Y and Gen Z Global Workplace Expectations Study," September 2, 2014, http://millennialbranding.com/2014/geny-genz-global-workplace-expectations-study/.

13. Victoria Bekiempis, "Nearly 1 in 5 Americans Suffers from Mental Illness Each Year," *Newsweek*, February 28, 2014, http://www.newsweek.com/nearly-1-5-americans-suffer-mental-illness-each-year-230608.
14. "2017 Employee Financial Wellness Survey," PwC, April 2017, https://www.pwc.com/us/en/industries/private-company-services/library/financial-well-being-retirement-survey.html.
15. Future Workplace and Virgin Pulse, "The Work Connectivity Study," to be published November 13, 2018, at http://workplacetrends.com/the-work-connectivity-study/.
16. Joshua Bjerke, "Inaugural Study Finds Employee Wellbeing a Strong Predictor of Performance," Recruiter, October 5, 2012, https://www.recruiter.com/i/inaugural-study-finds-employee-wellbeing-a-strong-predictor-of-performance/.
17. "Millennials Plan to Redefine the C-Suite, Says New American Express Survey," American Express, November 29, 2017, http://about.americanexpress.com/news/pr/2017/millennials-plan-to-redefine-csuite-says-amex-survey.aspx.
18. D. Kahneman and A. Deaton, "High Income Improves Evaluation of Life but Not Emotional Well-Being," *Proceedings of the National Academy of Sciences* 107, no. 38 (2010): 16489–16493, https://doi.org/10.1073/pnas.1011492107.
19. Kerry Hannon, "People with Pals at Work More Satisfied, Productive," *USA Today*, August 13, 2006, http://usatoday30.usatoday.com/money/books/reviews/2006-08-13-vital-friends_x.htm.
20. Jennifer Robinson, "Well-Being Is Contagious (for Better or Worse)," Gallup, November 27, 2012, www.gallup.com/businessjournal/158732/wellbeing-contagious-better-worse.aspx.
21. Shannon Greenwood, "In 2017, Two-Thirds of U.S. Adults Get News from Social Media," Pew Research Center's Journalism Project, September 7, 2017, http://www.journalism.org/2017/09/07/news-use-across-social-media-platforms-2017/pi_17-08-23_socialmediaupdate_0-01/.
22. Victoria Ward, "Facebook Makes Us More Narrow-Minded, Study Finds," *Telegraph*, January 7, 2016, https://www.telegraph.co.uk/news/newstopics/howaboutthat/12086281/Facebook-makes-us-more-narrow-minded-study-finds.html.
23. Future Workplace and Virgin Pulse, "The Work Connectivity Study."
24. Dan Schawbel, "Richard Branson: His Views on Entrepreneurship, Well-Being and Work Friendships," *Forbes*, October 23, 2017, https://www.forbes.com/sites/danschawbel/2017/10/23/richard-branson-his-views-on-entrepreneurship-well-being-and-work-friendships/#68e4165755d2.
25. Kat Boogaard, "Instead of Work-Life Balance, Try to Achieve Work-Life Integration," Inc., August 15, 2016, https://www.inc.com/kat-boogaard/4-tips-to-better-integrate-your-work-and-life.html.

Chapter 2: Optimize Your Productivity

1. Dan Schawbel, "Steve Harvey: His Biggest Obstacles, Time Management and Best Career Advice," *Forbes*, December 18, 2012, https://www.forbes.com

/sites/danschawbel/2012/12/18/steve-harvey-his-biggest-obstacles-time-management-and-best-advice/#49b754dd442a.

2. Kronos Inc. and Future Workplace, "The Employee Burnout Crisis: Study Reveals Big Workplace Challenge in 2017," January 9, 2017, https://www.kronos.com/about-us/newsroom/employee-burnout-crisis-study-reveals-big-workplace-challenge-2017.
3. Ian Hardy, "Losing Focus: Why Tech Is Getting in the Way of Work," *BBC News*, May 8, 2015, http://www.bbc.com/news/business-32628753.
4. Virgin Pulse, "95% of Employees Are Distracted During the Workday, New Virgin Pulse Survey Finds," October 22, 2014, https://www.virginpulse.com/press/95-of-employees-are-distracted-during-the-workday-new-virgin-pulse-survey-finds/.
5. Harmon.ie, "Collaboration & Social Tools Drive Business Productivity, Costing Millions in Work Interruptions," May 18, 2011, https://harmon.ie/press-releases/collaboration-social-tools-drain-business-productivity-costing-millions-work.
6. Vanessa K. Bohns, "A Face-to-Face Request Is 34 Times More Successful Than an Email," *Harvard Business Review*, April 11, 2017, https://hbr.org/2017/04/a-face-to-face-request-is-34-times-more-successful-than-an-email.
7. Staples Business Advantage, "The North American Workplace Survey," June 29, 2015, http://workplacetrends.com/north-american-workplace-survey/.
8. Future Workplace and Virgin Pulse, "The Work Connectivity Study," to be published November 13, 2018, at http://workplacetrends.com/the-work-connectivity-study/.
9. Joe Myers, "Is Technology Making Us Less Productive?," World Economic Forum, March 7, 2016, https://www.weforum.org/agenda/2016/03/is-technology-making-us-less-productive/.
10. Future Workplace and Beyond.com, "The Multi-Generational Leadership Study," November 10, 2015, http://workplacetrends.com/the-multi-generational-leadership-study/.
11. Dan Schawbel, "Charles Duhigg: How to Become More Productive in the Workplace," *Forbes*, July 24, 2016, https://www.forbes.com/sites/danschawbel/2016/07/24/charles-duhigg-how-to-become-more-productive-in-the-workplace/#536731c36d36.
12. Nicholas Bloom, "To Raise Productivity, Let More Employees Work from Home," *Harvard Business Review*, January 1, 2014, https://hbr.org/2014/01/to-raise-productivity-let-more-employees-work-from-home.
13. Future Workplace and Polycom, "The Human Face of Remote Working Study," March 21, 2017, http://workplacetrends.com/the-human-face-of-remote-working-study/.
14. Staples Business Advantage, "The North American Workplace Survey."
15. Jason Bramwell, "What Day Is the Most Productive? Tuesday!," AccountingWEB, December 23, 2013, https://www.accountingweb.com/practice/growth/what-day-is-the-most-productive-tuesday.
16. National Sleep Foundation, "How Much Sleep Do We Really Need?," n.d., https://sleepfoundation.org/how-sleep-works/how-much-sleep-do-we-really-need/page/0/2.

17. Lisa Evans, "The Exact Amount of Time You Should Work Every Day," *Fast Company*, September 15, 2014, https://www.fastcompany.com/3035605/the-exact-amount-of-time-you-should-work-every-day.
18. Dave Mielach, "Exercise Is Good for Your Health and Your Career," *Business News Daily*, February 24, 2012, https://www.businessnewsdaily.com/2084-exercise-good-health-career.html.
19. Christian Nordqvist, "Calories: Recommended Intake, Burning Calories, Tips, and Daily Needs," *MedicalNewsToday*, February 12, 2018, www.medicalnewstoday.com/articles/245588.php.
20. David T. Neal et al., "Habits—A Repeat Performance," *Current Directions in Psychological Science* 15, no. 4 (2006): 198–202, doi:10.1111/j.1467-8721.2006.00435.x.
21. "Global Mobile Consumer Survey: US Edition," Deloitte United States, February 28, 2018, https://www2.deloitte.com/us/en/pages/technology-media-and-telecommunications/articles/global-mobile-consumer-survey-us-edition.html.
22. Julie C. Bowker et al., "How BIS/BAS and Psycho-Behavioral Variables Distinguish Between Social Withdrawal Subtypes During Emerging Adulthood," *Personality and Individual Differences* 119 (2017): 283–288, doi:10.1016/j.paid.2017.07.043.
23. May Wong, "Stanford Study Finds Walking Improves Creativity," Stanford News, April 24, 2014, https://news.stanford.edu/2014/04/24/walking-vs-sitting-042414/.
24. Workfront, "2016–2017 State of Enterprise Work Report: U.S. Edition," July 1, 2016, https://resources.workfront.com/workfront-awareness/2016-state-of-enterprise-work-report-u-s-edition.

Chapter 3: Practice Shared Learning

1. Dan Schawbel, "Trevor Noah: Growing Up with Trauma, Being an Immigrant and His Views on the Election," *Forbes*, November 15, 2016, https://www.forbes.com/sites/danschawbel/2016/11/15/trevor-noah-growing-up-with-trauma-being-an-immigrant-and-his-views-on-the-election/#a07b3ae3b4c5.
2. Anuradha A. Gokhale, "Collaborative Learning Enhances Critical Thinking," *Journal of Technology Education* 7, no. 1 (1995), https://doi.org/10.21061/jte.v7i1.a.2.
3. Cornerstone OnDemand, "New Study Shows Who Sits Where at Work Can Impact Employee Performance and Company Profits," Cornerstone, July 27, 2016, https://www.cornerstoneondemand.com/company/news/press-releases/new-study-shows-who-sits-where-work-can-impact-employee-performance-and-company.
4. Kronos and WorkplaceTrends, "The Corporate Culture and Boomerang Employee Study," September 1, 2015, http://workplacetrends.com/the-corporate-culture-and-boomerang-employee-study/.

Chapter 4: Promote Diverse Ideas

1. Dan Schawbel, "Ed Catmull: What You Can Learn About Creativity from Pixar," *Forbes*, April 8, 2014, https://www.forbes.com/sites/dans

chawbel/2014/04/08/ed-catmull-what-you-can-learn-about-creativity-from-pixar/#4460ac3f4222.

2. Jessica Guynn et al., "Few Minorities in Non-Tech Jobs in Silicon Valley, USA TODAY Finds," *USA Today*, December 29, 2014, http://www.usatoday.com/story/tech/2014/12/29/usa-today-analysis-finds-minorities-under represented-in-non-tech-tech-jobs/20868353/.
3. Grace Donnelly, "Tech Employees Overestimate How Well Their Companies Promote Diversity," *Fortune*, March 22, 2017, fortune.com/2017/03/22/tech-employees-overestimate-how-well-their-companies-promote-diversity.
4. Catalyst, "Women in Management," February 7, 2017, http://www.catalyst.org/knowledge/women-management.
5. John Tierney, "Will You Be E-Mailing This Column? It's Awesome," *New York Times*, February 8, 2010, http://www.nytimes.com/2010/02/09/science/09tier.html.
6. Richard Fry, "Millennials Projected to Overtake Baby Boomers as America's Largest Generation," Pew Research Center, March 1, 2018, http://www.pewresearch.org/fact-tank/2018/03/01/millennials-overtake-baby-boomers/.
7. US Census Bureau, "Educational Attainment in the United States: 2017," December 14, 2017, https://www.census.gov/data/tables/2017/demo/education-attainment/cps-detailed-tables.html.
8. Laura Pappano, "The Master's as the New Bachelor's," *New York Times*, July 22, 2011, http://www.nytimes.com/2011/07/24/education/edlife/edl-24masters-t.html.
9. William Boston, "Bad News? What Bad News? Volkswagen Bullish Despite Emissions Costs," *Wall Street Journal*, April 28, 2016, https://www.wsj.com/articles/volkswagen-says-diesel-car-buy-backs-to-cost-almost-9-billion-1461831943.
10. Christiaan Hetzner, "VW Ex-Chairman Piech Challenges Board Nominees, Report Says," *Automotive News*, April 30, 2015, http://www.autonews.com/article/20150430/COPY01/304309944/vw-ex-chairman-piech-challenges-board-nominees-report-says.
11. Lea Winerman, "E-Mails and Egos," PsycEXTRA Dataset, February 2006, http://www.apa.org/monitor/feb06/egos.aspx.
12. Vanessa K. Bohns, "A Face-to-Face Request Is 34 Times More Successful Than an Email," *Harvard Business Review*, April 11, 2017, https://hbr.org/2017/04/a-face-to-face-request-is-34-times-more-successful-than-an-email.
13. Steven Pressfield, "Writing Wednesdays: Resistance and Self-Loathing," November 6, 2013, https://www.stevenpressfield.com/2013/11/resistance-and-self-loathing/.
14. Korn Ferry, "Executive Survey Finds a Lack of Focus on Diversity and Inclusion Key Factor in Employee Turnover," March 2, 2015, https://www.kornferry.com/press/executive-survey-finds-a-lack-of-focus-on-diversity-and-inclusion-key-factor-in-employee-turnover/.
15. Charles Duhigg, "What Google Learned from Its Quest to Build the Perfect Team," *New York Times*, February 25, 2016, https://www.nytimes.com/2016/02/28/magazine/what-google-learned-from-its-quest-to-build-the-perfect-team.html.

16. Dan Schawbel, "Adam Grant: Why You Shouldn't Hire for Cultural Fit," *Forbes*, February 2, 2016, https://www.forbes.com/sites/danschawbel/2016/02/02/adam-grant-why-you-shouldnt-hire-for-cultural-fit/#58d045717eba.
17. Future Workplace and Beyond, "The Multi-Generational Leadership Study," November 19, 2015, http://workplacetrends.com/the-multi-generational-leadership-study/.

Chapter 5: Embrace Open Collaboration

1. Dan Schawbel, "Beth Comstock: Being an Introverted Leader in an Extroverted Business World," *Forbes*, October 20, 2016, https://www.forbes.com/sites/danschawbel/2016/10/20/beth-comstock-being-an-introverted-leader-in-an-extroverted-business-world/#5d544eaa594f.
2. Lydia Saad, "The '40-Hour' Workweek Is Actually Longer—by Seven Hours," August 29, 2014, http://news.gallup.com/poll/175286/hour-work week-actually-longer-seven-hours.aspx.
3. Christine Congdon et al., "Balancing 'We' and 'Me': The Best Collaborative Spaces Also Support Solitude," *Harvard Business Review*, October 1, 2014, https://hbr.org/2014/10/balancing-we-and-me-the-best-collaborative-spaces-also-support-solitude.
4. Future Workplace and Randstad, "Despite the Tech Revolution, Gen Z and Millennials Crave In-Person Collaboration," September 6, 2016, https://www.randstadusa.com/about/news/despite-the-tech-revolution-gen-z-and-millennials-crave-in-person-collaboration/.
5. Lynda Gratton and Tamara J. Erickson, "Eight Ways to Build Collaborative Teams," *Harvard Business Review*, November 1, 2007, https://hbr.org/2007/11/eight-ways-to-build-collaborative-teams.
6. Jerry Useem, "When Working from Home Doesn't Work," *The Atlantic*, November 1, 2017, https://www.theatlantic.com/magazine/archive/2017/11/when-working-from-home-doesnt-work/540660/.
7. Steven Levy, "Apple's New Campus: An Exclusive Look Inside the Mothership," *Wired*, May 16, 2017, https://www.wired.com/2017/05/apple-park-new-silicon-valley-campus/.
8. Meeting King, "$37 Billion per Year in Unnecessary Meetings, What Is Your Share?," October 21, 2013, https://meetingking.com/37-billion-per-year-unnecessary-meetings-share/.
9. Patricia Reaney, "U.S. Workers Spend 6.3 Hours a Day Checking Email: Survey," *Huffington Post*, May 13, 2016, https://www.huffingtonpost.com/entry/check-work-email-hours-survey_us_55ddd168e4b0a40aa3ace672.
10. Kenneth Burke, "How Many Texts Do People Send Every Day?," Text Request, May 18, 2016, https://www.textrequest.com/blog/many-texts-people-send-per-day/.
11. Vanessa K. Bohns, "A Face-to-Face Request Is 34 Times More Successful Than an Email," *Harvard Business Review*, April 11, 2017, https://hbr.org/2017/04/a-face-to-face-request-is-34-times-more-successful-than-an-email.
12. Future Workplace and Randstad, "Despite the Tech Revolution, Gen Z and Millennials Crave In-Person Collaboration."

Chapter 6: Reward Through Recognition

1. Dan Schawbel, "Gary Vaynerchuk: Managers Should Be Working for Their Employees," *Forbes*, March 8, 2016, https://www.forbes.com/sites/danschawbel/2016/03/08/gary-vaynerchuk-managers-should-be-working-for-their-employees/#4e9df3e92008.
2. Dan Schawbel, "David Novak: Why Recognition Matters in the Workplace," *Forbes*, May 23, 2016, https://www.forbes.com/sites/danschawbel/2016/05/23/david-novak-why-recognition-matters-in-the-workplace/#58354e497bb4.
3. The Maritz Institute, "The Human Science of Giving Recognition," Maritz White Paper, January 2011, http://www.maritz.com/~/media/Files/MaritzDotCom/White%20Papers/Motivation/White_Paper_The_Science_of_Giving_Recognition1.pdf.
4. Maritz Institute, "Human Science of Giving Recognition."
5. E4S, "Incentives Bring Loyalty, Says Survey," June 7, 2008, http://www.e4s.co.uk/news/articles/view/747/job-news-and-information/gap-temp/Incentives-bring-loyalty-says-survey.
6. Daniel H. Pink, *Drive: The Surprising Truth About What Motivates Us* (New York: Riverhead Books, 2012).
7. Sho K. Sugawara, Satoshi Tanaka, Shuntaro Okazaki, Katsumi Watanabe, and Norihiro Sadato, "Social Rewards Enhance Offline Improvements in Motor Skill," *PLoS ONE* 7, no. 11 (2012): e48174, https://doi: 10.1371/journal.pone.0048174.
8. Badgeville, "Study on Employee Engagement Finds 70% of Workers Don't Need Monetary Rewards to Feel Motivated," June 13, 2013, https://badgeville.com/study-on-employee-engagement-finds-70-of-workers-dont-need-monetary-rewards-to-feel-motivated-211394831.html.
9. Melissa Dahl, "How to Motivate Your Employees: Give Them Compliments and Pizza," The Cut (blog), *New York Magazine*, August 29, 2016, https://www.thecut.com/2016/08/how-to-motivate-employees-give-them-compliments-and-pizza.html.
10. Dan Ariely and Matt R. Trower, *Payoff: The Hidden Logic That Shapes Our Motivations* (New York: TED Books/Simon & Schuster, 2016).
11. Martin Dewhurst et al., "Motivating People: Getting Beyond Money," *McKinsey Quarterly* (November 2009), https://www.mckinsey.com/business-functions/organization/our-insights/motivating-people-getting-beyond-money.
12. Future Workplace and Virgin Pulse, "The Work Connectivity Study," to be published November 13, 2018, at http://workplacetrends.com/the-work-connectivity-study/.
13. Shawn Bakker, "A Study of Employee Engagement in the Canadian Workplace," Psychometrics Canada, n.d., https://www.psychometrics.com/knowledge-centre/research/engagement-study/.
14. WorldatWork, "Trends in Employee Recognition," June 1, 2013, https://www.worldatwork.org/docs/research-and-surveys/Survey-Brief-Trends-in-Employee-Recognition-2013.pdf.

15. US Bureau of Labor Statistics, "Employee Tenure Summary," September 22, 2016, https://www.bls.gov/news.release/tenure.nr0.htm.
16. US Bureau of Labor Statistics, "Employee Tenure Summary."
17. Martin Berman-Gorvine, "Employee Peer Recognition Boosts Work Engagement," Bloomberg, December 19, 2016, https://www.bna.com/employee-peer-recognition-n73014448785/.
18. Francesca Gino and Adam Grant, "The Big Benefits of a Little Thanks," *Harvard Business Review*, March 30, 2015, https://hbr.org/2013/11/the-big-benefits-of-a-little-thanks.
19. Emiliana R. Simon-Thomas and Jeremy Adam Smith, "How Grateful Are Americans?," *Greater Good Magazine*, January 10, 2013, https://greatergood.berkeley.edu/article/item/how_grateful_are_americans.
20. Harvard Health Publishing, "In Praise of Gratitude," *Harvard Mental Health Letter*, November 2011, https://www.health.harvard.edu/newsletter_article/in-praise-of-gratitude.
21. Erin Holaday Ziegler, "Gratitude as an Antidote to Aggression," College of Arts & Sciences, University of Kentucky, October 20, 2011, https://psychology.as.uky.edu/gratitude-antidote-aggression.
22. Gino and Grant, "The Big Benefits of a Little Thanks."

Chapter 7: Hire for Personality

1. Dan Schawbel, "Richard Branson's Three Most Important Leadership Principles," *Forbes*, September 23, 2014, https://www.forbes.com/sites/danschawbel/2014/09/23/richard-branson-his-3-most-important-leadership-principles/#b7801e63d509.
2. Randstad US, "An Over-Automated Recruitment Process Leaves Candidates Frustrated and Missing Personal Connections, Finds Randstad US Study," August 3, 2017, https://www.randstadusa.com/about/news/an-over-automated-recruitment-process-leaves-candidates-frustrated-and-missing-personal-connections-finds-randstad-us-study/.
3. Inc., "Tony Hsieh: 'Hiring Mistakes Cost Zappos.com $100 Million,'" November 15, 2012, https://www.youtube.com/watch?v=XHcyKU-wZoA&feature=youtu.be.
4. Newton Software, "The Real Cost of a Bad Hire," July 6, 2016, https://newtonsoftware.com/blog/2016/07/06/the-real-cost-of-a-bad-hire/.
5. Millennial Branding and Beyond.com, "The Cost of Millennial Retention Study," December 6, 2013, http://millennialbranding.com/2013/cost-millennial-retention-study/.
6. CareerBuilder, "Nearly Seven in Ten Businesses Affected by a Bad Hire in the Past Year, According to CareerBuilder Survey," December 13, 2012, http://www.careerbuilder.com/share/aboutus/pressreleasesdetail.aspx?sd=12/13/2012&id=pr730&ed=12/31/2012.
7. Future Workplace and Virgin Pulse, "The Work Connectivity Study," to be published November 13, 2018, at http://workplacetrends.com/the-work-connectivity-study/.
8. Simon Chandler, "The AI Chatbot Will Hire You Now," *Wired*, September 13, 2017, https://www.wired.com/story/the-ai-chatbot-will-hire-you-now.

9. Chad Brooks, "Skip Skype: Why Video Job Interviews Are Bad for Everyone," *Business News Daily*, July 30, 2013, https://www.businessnewsdaily.com/4834-video-skype-job-interview.html.
10. Nikki Blacksmith et al., "Technology in the Employment Interview: A Meta-Analysis and Future Research Agenda," *Personnel Assessment and Decisions* 2, no. 1 (2016), doi:10.25035/pad.2016.002.
11. Millennial Branding and Beyond.com, "The Multi-Generational Job Search Study," May 20, 2014, http://millennialbranding.com/2014/multi-generational-job-search-study-2014/.
12. Tracy Moore, "When Headcount Is in Stride with Revenue (the Right Balance)," LinkedIn, June 4, 2015, https://www.linkedin.com/pulse/right-balance-when-headcount-stride-revenue-tracy-moore/.
13. Jessica Stillman, "8 Powerful Lessons You Can Learn from the Career of Elon Musk," *Inc.*, August 18, 2016, https://www.inc.com/jessica-stillman/8-powerful-lessons-you-can-learn-from-the-career-of-elon-musk.html.
14. Chris Anderson, "16 Management Quotes from the Top Managers in the World," Smart Business Trends, May 20, 2013, http://smartbusinesstrends.com/16-management-quotes/.
15. Lauren A. Rivera, "Hiring as Cultural Matching," *American Sociological Review* 77, no. 6 (2012): 999–1022, https://doi.org/10.1177/0003122412463213.
16. Dan Schawbel, "Hire for Attitude," *Forbes*, January 23, 2012, https://www.forbes.com/sites/danschawbel/2012/01/23/89-of-new-hires-fail-because-of-their-attitude/#5959ffb2137a.
17. Matthew Hutson and Tori Rodriguez, "Dress for Success: How Clothes Influence Our Performance," *Scientific American*, January 1, 2016, https://www.scientificamerican.com/article/dress-for-success-how-clothes-influence-our-performance/.
18. Chad A. Higgins and Timothy A. Judge, "Effect of Applicant Influence Tactics on Recruiter Perceptions of Fit and Hiring Recommendations: A Field Study," PsycEXTRA Dataset, doi:10.1037/0021-9010.89.4.622.
19. Joel Goldstein, "6 Personality Traits Employers Look for When Hiring," Lifehack, April 25, 2014, https://www.lifehack.org/articles/work/6-personality-traits-employers-look-for-when-hiring.html.
20. Christine Marino, "7 Need-to-Know Facts About Employee Onboarding," HR.com, July 7, 2015, blog.clickboarding.com/7-need-to-know-facts-about-employee-onboarding.
21. Kelsie Davis, "3 Questions Your New Hire Will Have on the First Day," Bamboo Blog, BambooHR, August 26, 2014, https://www.bamboohr.com/blog/new-hire-first-day/.

Chapter 8: Engage to Retain

1. Schawbel, Dan. "Personal Branding Interview: Tom Rath." *Personal Branding Blog*, October 25 2009, www.personalbrandingblog.com/personal-branding-interview-tom-rath/.

2. CareerBuilder, "New CareerBuilder Study Unveils Surprising Must Knows for Job Seekers and Companies Looking to Hire," July 1, 2016, https://www.careerbuilder.com/share/aboutus/pressreleasesdetail.aspx?ed=12%2F31%2F2016&id=pr951&sd=6%2F1%2F2016.
3. Future Workplace and Virgin Pulse, "The Work Connectivity Study," to be published November 13, 2018, at http://workplacetrends.com/the-work-connectivity-study/.
4. Gallup, "State of the American Workplace Report," http://news.gallup.com/reports/199961/state-american-workplace-report-2017-aspx.
5. Staples Business Advantage, "The North American Workplace Survey," June 29, 2015, http://workplacetrends.com/north-american-workplace-survey/.
6. Future Workplace and Virgin Pulse, "The Work Connectivity Study."
7. Dee DePass, "Honeywell Ends Telecommuting Option," *Star Tribune*, October 21, 2016, http://www.startribune.com/honeywell-ends-telecommuting-option/397929641/.
8. Sabrina Parsons, "Marissa Meyer at Yahoo! Declares: Face Time Is the Key," *Forbes*, March 4, 2013, https://www.forbes.com/sites/sabrinaparsons/2013/03/04/marissa-meyer-at-yahoo-declares-face-time-is-the-key/#13b0f0112be9.
9. Will Oremus, "Now Meg Whitman Wants Everyone to Stop Working from Home, Too," *Slate Magazine*, October 8, 2013, http://www.slate.com/blogs/future_tense/2013/10/08/hp_working_from_home_ban_marissa_mayer_s_yahoo_policy_becomes_industry_narrative.html.
10. Lionel Valdellon, "Remote Work: Why Reddit and Yahoo! Banned It," Wrike, February 10, 2015, www.wrike.com/blog/remote-work-reddit-yahoo-banned/.
11. Jim Harter and Annamarie Mann, "The Right Culture: Not Just About Employee Satisfaction," *Business Journal*, April 12, 2017, Gallup, http://news.gallup.com/businessjournal/208487/right-culture-not-employee-happiness.aspx.
12. Santiago Jaramillo, "The Value of Employee Engagement," Emplify, July 2017, https://emplify.com/blog/the-value-of-employee-engagement/.
13. "Employee Engagement Levels Are Focus of Global Towers Perrin Study," *Monitor*, January 2006, http://www.keepem.com/doc_files/Towers_Perrin_0106.pdf.
14. Kim Elsbach et al., "How Passive 'Face Time' Affects Perceptions of Employees: Evidence of Spontaneous Trait Inference in Context," *SSRN Electronic Journal*, September 29, 2008, http://dx.doi.org/10.2139/ssrn.1295006.
15. Scott Turnquest, "Distributed Teams and Avoiding Face Time Bias," ThoughtWorks, April 24, 2013, www.thoughtworks.com/insights/blog/distributed-teams-and-avoiding-face-time-bias.
16. Dan Schawbel, "Amy Cuddy: How Leaders Can Be More Present in the Workplace," *Forbes*, February 16, 2016, https://www.forbes.com/sites/danschawbel/2016/02/16/amy-cuddy-how-leaders-can-be-more-present-in-the-workplace/#64de74c3731c.
17. Ken Sterling, "Why Mark Zuckerberg Thinks One-on-One Meetings Are the Best Way to Lead," *Inc.*, September 28, 2017, https://www.inc.com/ken

-sterling/why-mark-zuckerberg-thinks-one-on-one-meetings-are-best-way-to-lead.html.

18. Lorne Michaels Quotes, BrainyQuote, n.d., https://www.brainyquote.com/quotes/lorne_michaels_501975.
19. WorkplaceTrends.com and Virtuali, "The Millennial Leadership Survey," July 21, 2015, http://workplacetrends.com/the-millennial-leadership-survey/.
20. Adam Bryant, "In Sports or Business, Always Prepare for the Next Play," *New York Times*, November 10, 2012, http://www.nytimes.com/2012/11/11/business/jeff-weiner-of-linkedin-on-the-next-play-philosophy.html.
21. Dan Schawbel, "Drew Houston: Why the Most Successful Entrepreneurs Solve Big Problems," *Forbes*, May 23, 2017, https://www.forbes.com/sites/danschawbel/2017/05/23/drew-houston-why-the-most-successful-entrepreneurs-solve-big-problems/#6ed883f67acd.
22. Dan Schawbel, "Biz Stone: From His Mom's Basement to Cofounding Twitter," *Forbes*, April 1, 2014, https://www.forbes.com/sites/danschawbel/2014/04/01/biz-stone-from-his-moms-basement-to-co-founding-twitter/#137f51001161.
23. Shawn Achor, "The Happiness Dividend," *Harvard Business Review*, June 23, 2011, hbr.org/2011/06/the-happiness-dividend.
24. M. D. Lieberman and N. I. Eisenberger. "Pains and Pleasures of Social Life," *Science* 323, no. 5916 (2009): 890–891, doi:10.1126/science.1170008.
25. Dan Schawbel, "Michael E. Porter on Why Companies Must Address Social Issues," *Forbes*, October 9, 2012, https://www.forbes.com/sites/danschawbel/2012/10/09/michael-e-porter-on-why-companies-must-address-social-issues/#3d3fc24e468a.
26. Dan Schawbel, "Personal Branding Interview: Simon Sinek," Personal Branding Blog, February 15, 2010, http://www.personalbrandingblog.com/personal-branding-interview-simon-sinek/.
27. Paul J. Zak, "The Neuroscience of Trust," *Harvard Business Review* (January–February 2017), https://hbr.org/2017/01/the-neuroscience-of-trust.
28. Future Workplace and Virgin Pulse, "The Work Connectivity Study."
29. Millennial Branding and Randstad, "Gen Y and Gen Z Global Workplace Expectations Study," September 2, 2014, http://millennialbranding.com/2014/geny-genz-global-workplace-expectations-study/.
30. Keith Ferrazzi, "Getting Virtual Teams Right," *Harvard Business Review* (December 2014), https://hbr.org/2014/12/getting-virtual-teams-right.

Chapter 9: Lead with Empathy

1. Dan Schawbel, "David Ortiz: From a Dominican Upbringing to 3-Time World Series Champion," *Forbes*, May 16, 2017, https://www.forbes.com/sites/danschawbel/2017/05/16/david-ortiz-from-a-dominican-upbringing-to-3-time-world-series-champion/#6c7a11e877a8.
2. Quoted in Jennifer Oldham and Liz Willen, "Are Texting, Multitasking Teens Losing Empathy Skills? Some Differing Views," HechingerEd, June 10, 2011,

http://hechingered.org/content/are-texting-multitasking-teens-losing-empathy-skills-some-differing-views_4002/.

3. Drake Baer, "An MIT Researcher Found 2 Scary Things That Happen When You're on a Phone All Day," Business Insider, October 20, 2015, http://www.businessinsider.com/mit-researcher-sherry-turkle-says-phones-make-us-lose-empathy-2015-10.
4. "Maya Angelou > Quotes > Quotable Quote," Goodreads, n.d., https://www.goodreads.com/quotes/5934-i-ve-learned-that-people-will-forget-what-you-said-people.
5. "2006 Northwestern Commencement—Sen. Barack Obama," NorthwesternU, July 15, 2008, https://www.youtube.com/watch?v=2MhMRYQ9Ez8.
6. Jeff Cox, "CEOs Make 271 Times the Pay of Most Workers," CNBC, July 20, 2017, https://www.cnbc.com/2017/07/20/ceos-make-271-times-the-pay-of-most-workers.html.
7. Maya Kosoff, "Mass Firings at Uber as Sexual Harassment Scandal Grows," *Vanity Fair*, June 6, 2017, https://www.vanityfair.com/news/2017/06/uber-fires-20-employees-harassment-investigation.
8. Maeve Duggan, "Online Harassment 2017," Pew Research Center, July 11, 2017, http://www.pewinternet.org/2017/07/11/online-harassment-2017/.
9. "2017 WBI US Survey: Infographic of Major Workplace Bullying Findings," June 24, 2017, http://www.workplacebullying.org/tag/bullying-statistics/.
10. Dawn Giel, "Wells Fargo Fake Account Scandal May Be Bigger Than Thought," CNBC, May 12, 2017, https://www.cnbc.com/2017/05/12/wells-fargo-fake-account-scandal-may-be-bigger-than-thought.html.
11. Andre Lavoie, "How to Get Rid of Toxic Office Politics," Work Smart, *Fast Company*, April 10, 2014, https://www.fastcompany.com/3028856/how-to-make-office-politicking-a-lame-duck.
12. Jamil Zaki, "What, Me Care? Young Are Less Empathetic," *Scientific American*, January 1, 2011, https://www.scientificamerican.com/article/what-me-care/.
13. Justin Bariso, "This Email from Elon Musk to Tesla Employees Is a Master Class in Emotional Intelligence," *Inc.*, June 14, 2017, https://www.inc.com/justin-bariso/elon-musk-sent-an-extraordinary-email-to-employees-and-taught-a-major-lesson-in.html.
14. Almie Rose, "One Woman's Brave Email Is Helping to Break the Mental Health Stigma," attn:, July 10, 2017, https://www.attn.com/stories/18200/how-email-helping-break-mental-health-stigma.
15. Harry McCracken, "Satya Nadella Rewrites Microsoft's Code," *Fast Company*, September 18, 2017, https://www.fastcompany.com/40457458/satya-nadella-rewrites-microsofts-code.
16. Belinda Parmar, "Want to Be More Empathetic? Here's Some Advice from a Navy SEAL," World Economic Forum, December 13, 2016, https://www.weforum.org/agenda/2016/12/what-a-navy-seal-can-teach-business-leaders-about-empathy/.

17. Anne Loehr, "Seven Practical Tips for Increasing Empathy," Blog, April 7, 2016, http://www.anneloehr.com/2016/04/07/increasing-empathy/.
18. "Patient Photos Spur Radiologist Empathy and Eye for Detail," RSNA Press Release, December 2, 2008, http://press.rsna.org/timssnet/media/pressreleases/pr_target.cfm?ID=389.
19. Adam M. Grant, "The Significance of Task Significance: Job Performance Effects, Relational Mechanisms, and Boundary Conditions," *Journal of Applied Psychology* 93, no. 1 (2008): 108–124, doi:10.1037/0021-9010.93.1.108.
20. Craig Dowden, "Forget Ethics Training: Focus on Empathy," *Financial Post*, February 27, 2015, http://business.financialpost.com/executive/c-suite/forget-ethics-training-focus-on-empathy.
21. Businessolver.com, "Empathy at Work: Why Empathy Matters in the Workplace," Businessolver, n.d., https://www.businessolver.com/executive_summary#gref.
22. Shalini Misra, "New Study Shows Putting Cell Phones out of Sight Can Enhance In-Person Conversations," Virginia Tech, August 7, 2014, https://vtnews.vt.edu/articles/2014/08/080714-ncr-misrasmartphonestudy.html.
23. P. Mulder, "Communication Model by Albert Mehrabian," ToolsHero, 2012, https://www.toolshero.com/communication-skills/communication-model-mehrabian/.
24. William A. Gentry et al., "Empathy in the Workplace," 2016, http://www.ccl.org/wp-content/uploads/2015/04/EmpathyInTheWorkplace.pdf.
25. Stephanie Zacharek et al., "TIME Person of the Year 2017: The Silence Breakers," *Time*, December 7, 2017, http://time.com/time-person-of-the-year-2017-silence-breakers/.
26. "BBC—Sexual Harassment in the Work Place 2017," ComRes, November 12, 2017, http://www.comresglobal.com/polls/bbc-sexual-harassment-in-the-work-place-2017/.
27. Karishma Vaswani, "The Costs of Sexual Harassment in the Asian Workplace," *BBC News*, December 13, 2017, http://www.bbc.com/news/business-42218053.
28. "The Reckoning: 2017 & Sexual Misconduct," Challenger, Gray & Christmas, Inc., February 2, 2018, http://www.challengergray.com/press/press-releases/reckoning-2017-sexual-misconduct.
29. Kristine Phillips, "Lawmaker Who Led #MeToo Push Accused of Firing Aide Who Wouldn't Play Spin the Bottle," *Washington Post*, February 20, 2018, https://www.washingtonpost.com/news/post-nation/wp/2018/02/19/lawmaker-who-led-metoo-push-invited-staffer-to-play-spin-the-bottle-complaint-says/?utm_term=.1edecbc20392.
30. Madeleine Aggeler, "Facebook and Google Employees Can Ask Each Other Out Once, but Only Once," The Cut (blog), *New York Magazine*, February 6, 2018, https://www.thecut.com/2018/02/facebook-employees-can-ask-each-other-out-once-but-only-once.html.
31. Sue Shellenbarger, "Is It OK for Your Boss to Hug Your Intern?," *Wall Street Journal*, February 13, 2018, https://www.wsj.com/articles/at-the-office-talking-about-sexual-harassment-is-still-tough-1518532200.

Chapter 10: Improve Employee Experiences

1. Dan Schawbel, "Stanley McChrystal: What the Army Can Teach You About Leadership," *Forbes*, July 13, 2015, https://www.forbes.com/sites/danschawbel/2015/07/13/stanley-mcchrystal-what-the-army-can-teach-you-about-leadership/#4c295ce972d5.
2. "Radical Innovation in Firms Across Nations: The Preeminence of Corporate Culture," *Journal of Marketing* (December 2008), http://faculty.london.edu/rchandy/innovationnations.pdf.
3. Dan Schawbel, "Peter W. Schutz on Becoming the CEO of Porsche," *Forbes*, August 24, 2012, https://www.forbes.com/sites/danschawbel/2012/08/24/peter-w-schutz-on-becoming-the-ceo-of-porsche/#653e257815f4.
4. Wikipedia, s.v. "Peter Schutz," accessed November 6, 2017, https://en.wikipedia.org/wiki/Peter_Schutz#cite_note-8.
5. Mitchell Hoffman et al., "The Value of Hiring Through Employee Referrals in Developed Countries," *IZA World of Labor*, June 2017, doi:10.15185/izawol.369.
6. Michael Housman, "Enemies to Allies: 6 Ways Employee Relationships Affect the Workplace," LinkedIn, April 22, 2015, https://www.linkedin.com/pulse/enemies-allies-6-ways-employee-relationships-affect-the-workplace/.
7. Craig Knight and S. Alexander Haslam, "The Relative Merits of Lean, Enriched, and Empowered Offices: An Experimental Examination of the Impact of Workspace Management Strategies on Well-Being and Productivity," *Journal of Experimental Psychology: Applied* 16, no. 2 (2010): 158–172, doi:10.1037/a0019292.
8. "The Importance of a Pleasant Workspace," Workplace Property, n.d., www.industrial-space-to-let.co.uk/the-importance-of-a-pleasant-workspace.html.
9. Rose Hoare, "Are Cool Offices the Key to Success?," CNN, August 10, 2012, https://www.cnn.com/2012/08/10/business/global-office-coolest-offices/index.html.
10. Future Workplace and Beyond.com, "The Active Job Seeker Dilemma Study," April 19, 2016, http://workplacetrends.com/the-active-job-seeker-dilemma-study/.
11. Josh Bersin et al., "The Employee Experience: Culture, Engagement, and Beyond," Deloitte Insights, February 28, 2017, https://www2.deloitte.com/insights/us/en/focus/human-capital-trends/2017/improving-the-employee-experience-culture-engagement.html.
12. IBM and Globoforce, "The Employee Experience Index," October 4, 2016, http://www.globoforce.com/wp-content/uploads/2016/10/The_Employee_Experience_Index.pdf.
13. Future Workplace and Beyond.com, "The Active Job Seeker Dilemma Study."
14. "Retreats Build Teams, but Only 20% of Companies Use Them," CPA Practice Advisor, July 26, 2017, http://www.cpapracticeadvisor.com/news/12354785/retreats-build-teams-but-only-20-of-companies-use-them.

Conclusion: Become More Human

1. Dan Schawbel, "Dr. Oz: What He's Learned from Over a Decade in the Spotlight," *Forbes*, September 18, 2017, https://www.forbes.com/sites

/danschawbel/2017/09/18/dr-oz-what-hes-learned-from-over-a-decade-in-the-spotlight/#438e9a336c5b.

2. "An Open Letter: Research Priorities for Robust and Beneficial Artificial Intelligence," Future of Life Institute, January 2015, https://futureoflife.org/ai-open-letter.
3. "Tim Cook's MIT Commencement Address 2017," MIT, June 9, 2017, https://www.youtube.com/watch?v=ckjkz8zuMMs.
4. "Facebook CEO Mark Zuckerberg Delivers Harvard Commencement Full Speech," Global News, May 25, 2017, https://www.youtube.com/watch?v=4VwElW7SbLA.
5. Future Workplace and Virgin Pulse, "The Work Connectivity Study," to be published November 13, 2018, at http://workplacetrends.com/the-work-connectivity-study/.
6. Future Workplace and Konica Minolta, "The Workplace of the Future Study," November 29, 2016, http://workplacetrends.com/workplace-of-the-future-study/.
7. Tae Kim, "McDonald's Hits All-Time High as Wall Street Cheers Replacement of Cashiers with Kiosks," CNBC, June 22, 2017, https://www.cnbc.com/2017/06/20/mcdonalds-hits-all-time-high-as-wall-street-cheers-replacement-of-cashiers-with-kiosks.html.
8. Jeremy Kahn, "Domino's Will Begin Using Robots to Deliver Pizzas in Europe," Bloomberg, March 29, 2017, https://www.bloomberg.com/news/articles/2017-03-29/domino-s-will-begin-using-robots-to-deliver-pizzas-in-europe.
9. Rachael King, "Newest Workers for Lowe's: Robots," *Wall Street Journal*, October 28, 2014, https://www.wsj.com/articles/newest-workers-for-lowes-robots-1414468866.
10. John Markoff, "'Beep' Says the Bellhop," *New York Times*, August 11, 2014, https://www.nytimes.com/2014/08/12/technology/hotel-to-begin-testing-botlr-a-robotic-bellhop.html.
11. Neil Connor, "Legal Robots Deployed in China to Help Decide Thousands of Cases," *Telegraph*, August 4, 2017, https://www.telegraph.co.uk/news/2017/08/04/legal-robots-deployed-china-help-decide-thousands-cases/.
12. Sophia Yan, "A.I. Will Replace Half of All Jobs in the Next Decade, Says Widely Followed Technologist," CNBC, April 27, 2017, https://www.cnbc.com/2017/04/27/kai-fu-lee-robots-will-replace-half-of-all-jobs.html.
13. Alex Gray, "Goodbye, Maths and English. Hello, Teamwork and Communication?," World Economic Forum, February 16, 2017, https://www.weforum.org/agenda/2017/02/employers-are-going-soft-the-skills-companies-are-looking-for/.
14. Dan Schawbel, "Geoff Colvin: Why Humans Will Triumph over Machines," *Forbes*, August 4, 2015, https://www.forbes.com/sites/danschawbel/2015/08/04/geoff-colvin-why-humans-will-triumph-over-machines/2/#134eb10b2b54.
15. Mike R. Morrison and Neal J. Roese, "Regrets and the Need to Belong," PsycEXTRA Dataset, n.d., http://journals.sagepub.com/doi/abs/10.1177/1948550611435137.

Index

About the Author

Dan Schawbel is a *New York Times* bestselling author, a partner and research director at Future Workplace, and the founder of both Millennial Branding and WorkplaceTrends.com. He is the bestselling author of two career books, *Promote Yourself* and *Me 2.0*. Through his companies he has conducted dozens of research studies and worked with major brands, including American Express, GE, Microsoft, Virgin, IBM, Coca-Cola, and Oracle. Dan has interviewed more than two thousand of the world's most successful people, including Warren Buffett, Anthony Bourdain, Jessica Alba, will.i.am, Michael Bloomberg, Chelsea Handler, Colin Powell, Sheryl Sandberg, and Arnold Schwarzenegger. In addition, he has written countless articles for *Forbes, Fortune, Time, The Economist, Quartz,* the World Economic Forum, the *Harvard Business Review, The Guardian,* and other publications that have, combined, generated more than fifteen million views. Considered "one of today's more dynamic young entrepreneurs" by *Inc.* magazine, Dan has been profiled or quoted in more than two thousand media outlets, such as NBC's *The Today Show*, Fox News's *Fox & Friends*, MSNBC's *Your Business, The Steve Harvey Show,* the *Wall Street Journal, People Magazine, Wired Magazine, GQ, The Economist*, and NPR. He has been recognized on several lists, including *Inc.* magazine's "30 Under 30," *Forbes Magazine*'s "30 Under 30," *Business Insider*'s "40 Under 40," *Businessweek*'s "20 Entrepreneurs You Should Follow," and as one of *Workforce Magazine*'s "Game Changers."

Connect with Dan Online

Web: DanSchawbel.com
Facebook.com/DanSchawbel
LinkedIn.com/In/DanSchawbel
Twitter.com/DanSchawbel
Instagram.com/DanSchawbel